ELEMENTARY PARTIAL
DIFFERENTIAL EQUATIONS

VNR NEW MATHEMATICS LIBRARY

under the general editorship of

J. V. ARMITAGE
Professor of Mathematics
University of Nottingham

N. CURLE
Professor of Applied Mathematics
University of St Andrews

The aim of this series is to provide a
reliable modern coverage of those mainstream
topics that form the core of mathematical
instruction in universities and comparable
institutions. Each book deals concisely with
a well-defined key area in pure or applied
mathematics or statistics. Many of the
volumes are intended not solely for students
of mathematics, but also for engineering and
science students whose training demands a
firm grounding in mathematical methods.

Elementary Partial
Differential Equations

by

R. J. GRIBBEN

VAN NOSTRAND REINHOLD COMPANY
New York - Cincinnati - Toronto - London - Melbourne

© R. J. Gribben 1975

ISBN 0 442 30055 7 cloth
ISBN 0 442 30056 5 paper

Published by Van Nostrand Reinhold
Company Limited 25–28 Buckingham Gate,
London SW1E 6LQ

Printed in Great Britain by William Clowes
& Sons Limited, London, Colchester and
Beccles

Contents

Preface

Most students of physics, engineering and applied science at a university or technical college encounter partial differential equations in their studies and the chief aim of this book is to provide for them a simple introduction to the subject. The properties of the three main classes of hyperbolic, parabolic and elliptic equations are illustrated by studying in some detail simple representatives of these classes, namely the one-dimensional wave equation, the one-dimensional diffusion equation and the two-dimensional Laplace's equation, respectively. In this way, the mathematics is reduced to a less sophisticated level and students are enabled to see clearly the similarities amongst, and differences between, these types of equation, not only in the behaviour of their solutions, but also in the sort of situation where they arise and the nature of the initial and boundary conditions required.

Partial differential equations are regarded throughout as mathematical models of physical phenomena and a wide variety of physical applications is considered. Indeed, in this vein, it is tempting to attach a second, literal meaning to the word 'elementary' in the title since the partial differential equations of the phenomena considered are obtained mainly from applying suitable physical laws to a small element of the system. Passage to the limit as the element becomes vanishingly small then produces the appropriate model equation. Wherever appropriate, opportunity is taken to relate a mathematical solution to the corresponding physical situation.

Thus it is hoped that mathematical simplicity, without undue attention to rigour, will make the book of particular appeal to non-specialist mathematicians and provide a useful introduction to the principles and methods for those going on to study more general equations. Applied mathematics students, who normally study more advanced techniques, may draw some benefit from the physical emphasis.

The lay-out of the book is straightforward. An introductory chapter provides the basic concepts and the definitions and results needed later in the text. Some previous acquaintance with partial differentiation and Fourier series is required. The remaining three chapters deal with the wave, diffusion and Laplace's equation in turn. In each case the equation is derived in appropriate physical contexts and its properties discussed, especially with regard to the behaviour of its characteristics and its corresponding difference equation. Finally, solutions are obtained using primarily the method of separation of variables which is particularly easy for students to grasp for the equations considered. Worked examples are included and exercises appended to each chapter.

The author acknowledges with grateful thanks the assistance and helpful comments of Dr. R. J. Cole, who read the manuscript, and Mr. A. R. Veitch, who checked the proofs.

CHAPTER 1

Some Basic Ideas

1.1 Mathematical Modelling and Partial Differential Equations

In order to be able to make useful statements about the manner in which a system, whether it be physical, economic, statistical or any other, behaves in practice the mathematician must first replace it by an idealized system or mathematical model. This process involves the introduction of symbols to represent those real quantities that are important in the phenomenon and the formulation of laws or hypotheses which govern the way in which these symbols are related to each other. Once translated into this symbolic, abstract form the problem can then be subjected to any of the mathematical techniques that might prove suitable and which have evolved over the years, or it may even provide motivation for developing new ones. In any event, such results that are obtained at this level are re-interpreted in terms of quantities that have physical (or economic or statistical) meaning in the original system and can be tested by experience. This may be acquired by deliberately designing and carrying out an experiment to test directly, in controlled conditions, some predicted behaviour of the system. Alternatively, in systems that are uncontrollable (at least by man!), such as in naturally-occurring phenomena, observations and measurements are made on the large-scale system itself outside the laboratory. In either case, for so long as the system behaves in a way that the model predicts it should, the latter is deemed to be a good one; otherwise it must be modified and improved.

Mathematical model building can be a most difficult task and is subject to two conflicting requirements. On the one hand it must be sufficiently complex that it takes account of all the important components of the system and on the other it has to be simple enough to be amenable to mathematical analysis. Examples of systems successfully treated by the use of mathematical models are countless: the model of the solar system based on Newton's laws of mechanics, later modified to take into account relativistic effects, the continuum model of gases, liquids, and solids and the statistical model used to describe the flow of rarefied gases and plasmas. Others are still being sought; for example, the current models used by applied mathematicians and engineers for describing the turbulent motion of fluids are not fully satisfactory and the related problem of finding a model of the moving atmosphere on which successful weather prediction can be based is still largely unsolved.

1

The simplest models that are encountered in dynamics involve one degree of freedom; that is, a single variable or coordinate is capable of describing uniquely the state of the system. As an obvious example (Fig. 1.1(a)) we can think of a simple pendulum where a bob of mass but no volume is suspended from a fixed point 0 by a string of length but no mass. Friction is neglected at the suspension and the bob oscillates in a vertical plane under gravity. A knowledge of, say, the angle θ made by the string with the downward vertical through 0 carries with it a knowledge of the position of any point of the string or the bob itself. Upon applying Newton's Law of Motion to the bob it is well known that θ is determined for all time from the ordinary differential equation,

$$\frac{d^2\theta}{dt^2} + \frac{g}{l}\sin\theta = 0, \tag{1.1}$$

where g is the gravitational acceleration and l is the length of the string, when sufficient information about the way the motion is started, the initial conditions, is included. In any real oscillating pendulum, of course, the bob has a definite size, the string has mass and there is friction at the suspension, but provided these are all small in some sense, the mathematical model described by equation (1.1) is adequate.

The process is easily extended so that systems of two degrees of freedom can be considered. Thus, for example, for a double, compound pendulum consisting of two smoothly-jointed rods, suspended at one end

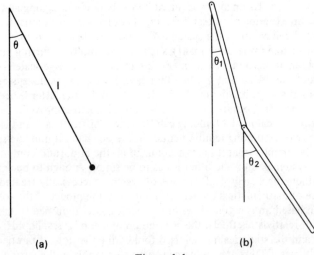

(a)   (b)

Figure 1.1

(Fig. 1.1(b)), a knowledge of two coordinates such as θ_1 and θ_2 for all time is necessary and sufficient to determine the position, velocity and acceleration of any point of the system at any time. According to the model, these coordinates are determined from a system of two coupled, ordinary differential equations for $\theta_1(t)$ and $\theta_2(t)$ and suitable initial conditions.

In general then, by continuing in this way, a mechanical system of n degrees of freedom is completely determined by a knowledge of exactly n suitable coordinates, $\theta_i(t)$ ($i = 1, \ldots, n$), which are called generalized coordinates. The $\theta_i(t)$ satisfy a coupled set of n ordinary differential equations and appropriate initial conditions.

Now consider the situation as n gets large and therefore so does the corresponding number of required coordinates. For a finite system in such circumstances the generalized coordinates can become congested as, for example, in the case of a horizontal, stretched, light, elastic string of length l fixed at its ends, A and B, and on which are fastened n small masses equally distributed, or a metal rod of length l on which are placed n equally-spaced thermometers to take temperature tappings. The former example is connected with the subject of Chapter 2 and the latter with that of Chapter 3. When the masses on the string move vertically, suitable generalized coordinates are their displacements u_i ($i = 1, \ldots, n$) from the equilibrium, horizontal position (ignoring the effects of gravity; see Fig. 1.2) and the number of coupled equations to solve for large n

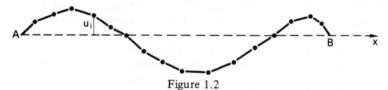

Figure 1.2

becomes large. At a given instant the masses define a polygonal arc between A and B (Fig. 1.2). For large n it becomes clear that we might view the matter in a slightly different mathematical light. We can regard the polygon as a close approximation to a smooth curve although we do not say precisely how the latter is to be constructed. The displacement of any point of this curve from AB (corresponding to a u_i) can be thought of, at a given t, as some value of a smooth function $u(x)$, where x is an additional space coordinate measuring distance from A along AB. Allowing for the variation in t the displacement function u is a single function of two variables, x and t. Thus, in this case, for a model of the system we might expect a single equation, which represents a gain, but involving partial derivatives, which introduces extra difficulties. Such an equation is called a partial differential equation. From the rough physical picture obtained by looking at Fig. 1.2 and imagining the number of attached particles to increase without limit we conjecture that ultimately, as the number of degrees of freedom tends to infinity, the connecting massless string disappears, the system loses its discrete character and we may consider the masses as pieces of a continuous, heavy, elastic string. The system becomes a continuum in this limiting case. In fact it serves as a model of an elastic string having a definite mass per unit length as described in Chapter 2.

In a similar way for the metal rod the temperature $u_i(t)$, which at the discrete level can be regarded as the uniform temperature of the small ith portion τ_i of length of rod, satisfying an appropriate equation of heat balance in τ_i, tends in the limit as n tends to infinity to a continuous function of t and distance along the rod x. Again the discrete view requires a solution of n simultaneous ordinary, differential equations for n unknowns, whilst the continuum approach needs the solution of one partial differential equation in two independent variables.

The rigorous process by which a discrete system is allowed to tend to a continuous one in the manner indicated above is fraught with difficulty, both philosophical and mathematical, and will not be our concern here. In the following chapters we shall generally assume a continuum to exist and apply physical laws to an elementary portion of it. Differential calculus then provides the apparatus for obtaining the partial differential equation for the system. In the final analysis, of course, irrespective of the method of derivation, a mathematical model of a physical system (of chief interest here) stands or falls by comparison with direct observation.

Equations which describe behaviour of continua are usually partial differential equations, three of the simplest of which are the objects of study in the ensuing chapters. The solution functions that are obtained are generally functions of position (as well as possibly of time) and are called vector or scalar fields as the case may be.

An important difference arises in the supplementary information that must be supplied to obtain unique solutions of ordinary differential equations on the one hand and partial differential equations on the other. Returning to the example of Fig. 1.2, the solution for the motion of the n masses is, in principle, uniquely determined if all the initial displacements and velocities are known since then the $2n$ arbitrary constants arising in the general solution of a system of n second order ordinary differential equations can be found. In the continuum limit, when the small masses, so to speak, make up the whole string (though its elastic properties are retained), it is plausible to say that knowing the initial displacement and velocity fields for $0 < x < l$ determines the subsequent motion of the string. These fields are, of course, functions of x and the statement is true provided constraints, stipulating that the ends $x = 0$ and $x = l$ are always fixed, are included. Such constraints are called boundary conditions and are discussed more fully in Section 4 below and the other chapters. Their counterparts in the discrete system appear as conditions on the direction of the forces acting along the first and last pieces of string (always towards A and B respectively) and as such would be included in the equations of motion for the first and last particles.

1.2 Linearity and Superposition

A partial differential equation then is simply an equation involving partial derivatives. Thus it must contain at least one dependent variable, for which we shall usually reserve the symbol u, and two independent variables. The specific equations we shall consider in Chapters 2, 3 and 4, respectively, are

(a) $\dfrac{\partial^2 u}{\partial x^2} - \dfrac{1}{c^2}\dfrac{\partial^2 u}{\partial y^2} = 0$,   (b) $\dfrac{\partial^2 u}{\partial x^2} - \dfrac{1}{\alpha^2}\dfrac{\partial u}{\partial y} = 0$,

(c) $\dfrac{\partial^2 u}{\partial x^2} + \dfrac{\partial^2 u}{\partial y^2} = 0$,

$$\tag{1.2}$$

where c and α are constant. These are all equations of second-order, by definition, because the order of the highest derivatives they contain is two. (In Chapters 2 and 3 the independent variable y is replaced by t, for time, on physical grounds.)

A more important property of equations (1.2), however, is that of *linearity*. In an equation, if all the terms which contain u or its derivatives are collected together and represented symbolically by the expression Lu, then the equation can be written in the form,

$$Lu = f(x, y), \tag{1.3}$$

where f is some given function of the independent variables x and y. The symbol L denotes the differential operator corresponding to the equation and it is linear, and, hence, so is equation (1.3), if the property,

$$L(c_1 u_1 + c_2 u_2) = c_1 Lu_1 + c_2 Lu_2, \tag{1.4}$$

for any two functions u_1 and u_2 and constants c_1 and c_2, is satisfied. For example, the operators $L \equiv \partial^2/\partial x^2 + \partial^2/\partial y^2$, $L \equiv g(x, y)\,\partial^2/\partial x\,\partial y$ (g given), are linear from application of the definition and the elementary rules of differentiation, whereas any operator involving powers (other than the first) of u or its derivatives, or products of u and its derivatives is nonlinear, e.g., $L \equiv u^2$, $L \equiv u\,\partial/\partial x$.

If $f \equiv 0$ in equation (1.3) it is homogeneous and the particular equations (1.2) are therefore both linear and homogeneous. An important property of such equations is that any two solutions can be used to generate further solutions. Thus, if u_1 and u_2 are solutions, from (1.4),

$$0 = c_1 Lu_1 + c_2 Lu_2 = L(c_1 u_1 + c_2 u_2),$$

for any two constants c_1 and c_2, and therefore $c_1 u_1 + c_2 u_2$ is also a solution.

It is clear that there is no need to stop at just two functions. Any n solutions $u_1, u_2, \ldots, u_n$ of a linear, homogeneous equation can be combined to form a solution $c_1 u_1 + \cdots + c_n u_n$, where the c_i ($i = 1, \ldots, n$) are arbitrary constants. This result constitutes the *Principle of Superposition* in its finite form. The proof follows easily from the definition (1.4) and an application of the method of induction.

It follows from the Principle of Superposition that a great many solutions of a linear, homogeneous differential equation can be written down, once several particular solutions are known. Leaving aside for the moment the question of boundary conditions, which will be discussed in Section 1.4, it is natural to enquire as to the possibility of adding up *infinite* sets of solutions to obtain more general ones. In other words, in what circumstances are we allowed to let $n \to \infty$ in the enunciation of the Principle of Superposition given above?

The answer is supplied by the Principle of Superposition *in its extended form* which deals with infinite combinations of solutions. It can be stated as follows.

The infinite sum,

$$u = \sum_{n=1}^{\infty} c_n u_n, \tag{1.5}$$

where c_n are constants, satisfies the linear partial differential equation, $Lu = 0$, provided that,

(i) $Lu_1 = 0, Lu_2 = 0, Lu_3 = 0, \ldots$
(ii) the series (1.5) is convergent and differentiable term-by-term as many times as needed in the definition of L.

We give a sketch of the proof. The series (1.5) must clearly converge in an appropriate domain of the independent variables. Furthermore, in order to perform the substitution into the differential equation it must be differentiated term-by-term. For example, a term in L such as

$$\frac{\partial u}{\partial x} = \frac{\partial}{\partial x} \left(\lim_{N \to \infty} \sum_{n=1}^{N} c_n u_n \right) = \lim_{N \to \infty} \left(\sum_{n=1}^{N} c_n \frac{\partial u_n}{\partial x} \right),$$

where N is an integer, and from the properties of infinite series this can be multiplied by any given function of x and y that might occur in the definition of L. Thus, more generally, condition (ii) means that Lu can be written as

$$L \left(\lim_{N \to \infty} \sum_{n=1}^{N} c_n u_n \right) = \lim_{N \to \infty} L \left(\sum_{n=1}^{N} c_n u_n \right).$$

But condition (i) and the Principle of Superposition in its finite form imply that $L(\sum_{n=1}^{N} c_n u_n) = 0$ for all finite N. Hence,

$$Lu = \lim_{N \to \infty} L \left(\sum_{n=1}^{N} c_n u_n \right) = 0.$$

The result established here is of the utmost importance in the theory of linear partial differential equations and we shall use it continually in the remaining chapters. Some further slight restriction is needed when we come to consider the behaviour of solutions at the boundaries and this will be mentioned in Section 1.4.

A way of ensuring that the series is differentiable in the domain of interest is to apply the properties of uniform convergence. A convergent series can be differentiated term-by-term if the terms of the differentiated series are continuous and this series itself converges uniformly. As an illustration we take the series (3.15),

$$u = \sum_{n=1}^{\infty} \frac{(-1)^{n+1}}{n} \sin\left(\frac{n\pi x}{l}\right) \exp\left(-\frac{\alpha^2 n^2 \pi^2 t}{l^2}\right),$$

$$t > 0, \quad 0 \leqslant x \leqslant l, \tag{1.6}$$

obtained in the solution of a physical problem in Chapter 3. Here, α and l are constants and x and t are the independent variables. Now the derivatives $\partial^2 u / \partial x^2$ and $\partial u / \partial t$ are contained in the diffusion equation of which (1.6) is a solution. Thus, formally, we obtain, for example, that

$$\frac{\partial^2 u}{\partial x^2} = \sum_{n=1}^{\infty} \left(\frac{\pi}{l}\right)^2 (-1)^n n \sin\left(\frac{n\pi x}{l}\right) \exp\left(-\frac{\alpha^2 n^2 \pi^2 t}{l^2}\right). \tag{1.7}$$

Now $|\sin(n\pi x/l)| \leqslant 1$, and if $t \geqslant T > 0$, $\exp(-\alpha^2 n^2 \pi^2 t/l^2) \leqslant \exp(-\alpha^2 n^2 \pi^2 T/l^2)$. Thus,

$$\left| \left(\frac{\pi}{l}\right)^2 (-1)^n n \sin\left(\frac{n\pi x}{l}\right) \exp\left(-\frac{\alpha^2 n^2 \pi^2 t}{l^2}\right) \right|$$

$$\leqslant \frac{\pi^2 n}{l^2} \exp\left(-\frac{\alpha^2 n^2 \pi^2 T}{l^2}\right) = d_n,$$

say, for $t \geqslant T$. The ratio,

$$\frac{d_{n+1}}{d_n} = \left(\frac{n+1}{n}\right) \exp\left\{-\frac{\alpha^2 \pi^2 T(2n+1)}{l^2}\right\} \to 0 \quad \text{as } n \to \infty,$$

and, hence, the series (1.7) for $\partial^2 u / \partial x^2$ is absolutely convergent by the comparison test. Moreover, it is uniformly convergent in $t \geqslant T, 0 \leqslant x \leqslant l$, by the Weierstrass M-test since the majorizing series Σd_n is a series of constants. The differentiation is therefore justified and, likewise, so is the term-by-term differentiation of (1.6) to form $\partial u / \partial t$.

However, it should be noted that where functions with discontinuities, or discontinuities in derivatives, are obtained as solutions, as arises often in the case of the wave equation in Chapter 2, then series representations of such solutions cannot be uniformly convergent in the whole domain of interest because the sum of a uniformly convergent series is continuous. Of course, at such discontinuities the differential equation itself is not defined.

1.3 Classification of Equations

If

$$Lu = 0, \tag{1.8}$$

is a second-order, linear, homogeneous partial differential equation in the two independent variables, x and y, the operator L must be of the form,

$$L \equiv a \frac{\partial^2}{\partial x^2} + 2h \frac{\partial^2}{\partial x \, \partial y} + b \frac{\partial^2}{\partial y^2} + l \frac{\partial}{\partial x} + m \frac{\partial}{\partial y} + n, \tag{1.9}$$

where we assume the coefficients a, h, b, l, m and n are given, continuously-differentiable functions of x and y. As with ordinary differential equations the highest derivatives in the equation govern its behaviour but now, with L given by (1.9), there are three possible such terms. It is found that the properties of the equation depend critically on the relative magnitudes of the coefficients a, h and b of these second-order derivatives. On the basis of this a classification of second-order equations can be made by which they are called

$$\left.\begin{array}{l}
\text{(a) hyperbolic if } ab - h^2 < 0, \\
\text{(b) parabolic if } ab - h^2 = 0, \\
\text{(c) elliptic if } ab - h^2 > 0.
\end{array}\right\} \tag{1.10}$$

We notice that the classification corresponds to whether the associated conic, $ax^2 + 2hxy + by^2 = 1$, represents a hyperbola, a parabola or an ellipse, respectively. Also, the conditions in (1.10a) and (1.10c) represent regions of the (x, y) plane so that in the general case an equation may well be of different type in different regions of the plane.

Equations (1.2), which we shall study in this book, provide the simplest, non-trivial examples of equations of each of the three types; (1.2a) is called the one-dimensional wave equation, (1.2b) is the one-dimensional diffusion equation and (1.2c) the two-dimensional Laplace's equation. The coefficients in these equations are all constant and it is easily verified, by applying the definition (1.10), that throughout the whole (x, y) plane the wave equation is hyperbolic, the diffusion equation is parabolic and Laplace's equation is elliptic. It is shown in the remaining chapters that these equations themselves arise as models for a wide variety of physical phenomena. Also, however, by a suitable transformation of the independent variables the general equation (1.8) with L given by (1.9) can be simplified so that the second-order derivatives take one or other of the forms of the second-order derivatives in (1.2) in the new variables. The equation is then said to be in its normal form. (Actually, in the hyperbolic case, the normal form is usually taken to mean that the second-order derivatives have the form $\partial^2 u/\partial x \, \partial y$, but (1.2a) can always be converted into this form as shown in Section 2.3.) It follows that a discussion of the simpler equations (1.2) is useful in describing the properties of the general equation (1.8) and (1.9), when we bear in mind that the second-order derivatives dominate its behaviour.

The equations are generally to be solved in some given region R in the space of the independent variables with definite boundary conditions to be satisfied on the boundary, or portion of the boundary, of R. In the ensuing chapters, as well as describing methods of solution of the three equations (1.2), we attempt to point out, in both mathematical and physical terms, the differences that arise in formulating problems and in the behaviour of the solutions. Such differences occur, for example, in the types of boundary conditions to be imposed (including initial conditions, when one independent variable is time t and a 'boundary' is at $t = 0$) to ensure a unique solution. This topic is discussed in Section 4 generally and, with special reference to their physical context, in the separate chapters. Questions of existence and uniqueness of solutions when suitable boundary conditions are applied are considered in Section 5.

There is a further important aspect of the classification into hyperbolic, parabolic and elliptic types. To see this we first note that the general second-order linear equation is equivalent to a system of three first-order simultaneous partial differential equations in three unknowns. For, if we write

$$p = \frac{\partial u}{\partial x}, \quad q = \frac{\partial u}{\partial y}, \tag{1.11}$$

then the equation defined by (1.8) and (1.9) becomes

$$a \frac{\partial p}{\partial x} + 2h \frac{\partial p}{\partial y} + b \frac{\partial q}{\partial y} + lp + mq + nu = 0, \tag{1.12}$$

which, together with (1.11), completes the system of three first-order equations. Furthermore, if (1.9) contains no term in u itself, as is the case with the three equations (1.2), the equivalent first-order system can be reduced to the two equations, (1.12), with $n = 0$, and

$$\frac{\partial p}{\partial y} - \frac{\partial q}{\partial x} = 0,$$

which is obtained when u is eliminated from equations (1.11). The reason for introducing equivalent first-order systems is to recognize more readily the possibility of obtaining curves, or families of curves, in the (x, y) plane having the property that the equations can be written in a form containing only total derivatives along these curves of particular, linear combinations of p and q. We can illustrate this by considering equation (1.12) with $l = m = n = 0$,

$$a \frac{\partial p}{\partial x} + 2h \frac{\partial p}{\partial y} + b \frac{\partial q}{\partial y} = 0. \tag{1.13}$$

and

$$\frac{\partial p}{\partial y} - \frac{\partial q}{\partial x} = 0. \tag{1.14}$$

On adding λ times equation (1.14) to (1.13), where λ is an arbitrary function of x and y, we obtain

$$a\left[\frac{\partial p}{\partial x} + \frac{(2h + \lambda)}{a}\frac{\partial p}{\partial y}\right] - \lambda\left[\frac{\partial q}{\partial x} - \frac{b}{\lambda}\frac{\partial q}{\partial y}\right] = 0. \tag{1.15}$$

Now the total derivative of a function $u(x, y)$ with respect to x along a curve $y = y(x)$ in the (x, y) plane is

$$\frac{d}{dx}\{u[x, y(x)]\} = \frac{\partial u}{\partial x} + \frac{\partial u}{\partial y}\frac{dy}{dx}. \tag{1.16}$$

Thus, in order to express (1.15) in the form,

$$a\frac{dp}{dx} - \lambda\frac{dq}{dx} = 0,$$

along curves $y = y(x)$, it is necessary to choose λ so that

$$\frac{(2h + \lambda)}{a} = -\frac{b}{\lambda} = \frac{dy}{dx}$$

or

$$\lambda^2 + 2h\lambda + ab = 0. \tag{1.17}$$

Since, from (1.15), the curves having the required property have slopes given by

$$\frac{dy}{dx} = \frac{2h + \lambda}{a}, \tag{1.18}$$

we see from (1.17) that two real families of such curves exist in the hyperbolic case, $ab - h^2 < 0$, one in the parabolic case, $ab - h^2 = 0$, and none in the elliptic case, $ab - h^2 > 0$. The curves defined by (1.18) are called the families of *characteristics*.

Characteristics have important properties; for example, discontinuities in second derivatives can occur only across characteristics (Exercise 1.5) and hence functions having such behaviour cannot be obtained as solutions of elliptic equations of type (1.8) and (1.9) (provided the coefficients are sufficiently smooth functions). The connection of characteristics with the particular equations (1.2) is discussed in the separate chapters and their important role in the numerical solution of hyperbolic equations is illustrated by reference to the wave equation in Section 2.4.

1.4 Initial and Boundary Conditions

In the field of ordinary differential equations a general solution can frequently be found which contains arbitrary constants; these are then determined from suitable boundary conditions applied on the solution at particular values of the independent variable. With partial differential equations the situation is similar in theory but rather more complex in practice. Firstly, the space of the independent variables is multidimensional although here the discussion is restricted to two independent variables. Hence we consider the solution of equations in regions, possibly infinite, of a plane. Secondly, it is much less probable that a general solution of a partial differential equation can be obtained and written down (even though such a step is possible for the wave equation in Chapter 2). Finally, such general solutions contain arbitrary functions and not constants.

As with ordinary differential equations this arbitrariness in the solution is removed by prescribing the behaviour at the boundaries. It is therefore important to know at the outset precisely what boundary information is needed; stated rather loosely, if too much boundary information is given the solution will not exist, and if too little, it will not be unique. Such information consists of not only the behaviour of the solution and/or its derivatives but also the location of those parts of the boundary at which values of these quantities are to be specified. Fortunately, in deriving partial differential equations to describe physical phenomena, as in the remaining chapters, the physics of the problem suggests boundary conditions that must be imposed. For example, in the case of plane, steady-state heat flow in a circular disc, discussed in Section 4.5.2, with the perimeter maintained at a given temperature by some external means, it is clear that the solution of the appropriate equation (in this case Laplace's equation) in the disc must approach the prescribed boundary value at the edge of the disc. Actually, it turns out that this constraint on the solution is sufficient to render it unique, provided that the restriction to functions with continuous derivatives is made.

Boundary conditions arise in two important forms when partial differential equations are applied to physical processes. In the first case some condition is imposed on the solution at all points of the boundary of the region in which the equation is to be solved. This is therefore called a *boundary value problem* and mainly occurs with the solution of elliptic equations as in the above example of steady-state heat flow in a disc. The most general form of the boundary condition that we shall require can be expressed in the operator form, $L_b u$ given on the boundary, where,

$$L_b = h + k \frac{d}{dn}$$
(1.19)

Here, h and k are constants and d/dn denotes the normal derivative. Special forms of this boundary condition are,

(a) $k = 0$. The solution u is prescribed on the boundary and the resulting problem is called a *Dirichlet problem*.

(b) $h = 0$. The normal derivative of the solution is prescribed on the boundary and the problem is then called a *Neumann problem*.

Problems in which h and k in (1.19) take different values at different parts of the boundary are also encountered in the text.

In describing the second form in which boundary conditions can arise it is convenient to single out one of the independent variables, which we denote by t as it is often in connection with time-varying phenomena that the condition arises. Then, if the equation is of nth order in t, values of the solution *and* its first $n - 1$ time derivatives are prescribed at $t = 0$ for all values of the other independent variable, say x. The problem is thus referred to as an *initial value problem*. An alternative phrase, *Cauchy problem*, is also used. It is not, in general, possible to prescribe the behaviour of the solution at some later value of t.

It often arises that the solution of a differential equation is required subject to conditions of Cauchy type on the boundary corresponding to one variable, usually on $t = 0$, and to conditions of Dirichlet or Neumann type on the boundary corresponding to the other, say at two finite values of x for the one-dimensional wave equation. Such a problem is called a *mixed problem*.

Initial value and mixed problems arise mainly in connection with parabolic and hyperbolic differential equations.

The distinction between Cauchy and Dirichlet problems is clearly indicated from considering the corresponding difference equations. In a Cauchy problem, associated with the wave or diffusion equation, a numerical solution can be advanced at successive time intervals starting from the initial instant (see Sections 2.4 and 3.4), but in a Dirichlet (or Neumann) problem, appropriate to Laplace's equation, such a solution requires a simultaneous calculation of values of the dependent variable in the whole region of interest (see Section 4.4).

The Principle of Superposition in its extended form can also be applied to any linear and homogeneous boundary condition written in the form $L_b u = 0$, where L_b is given by (1.19). The function $L_b u$, however, should then be continuous at the boundary.

1.5 Uniqueness of Solutions

We normally expect a physical system to respond in a unique way to a given set of conditions; the experimentalist likes his results to be reproducible. Thus any point of an elastic string, which is released from rest after being pulled into some initial shape, has the same variation of displacement (distance from the straight-line position) with time however often the experiment is repeated and in the example quoted in Section 4 the temperature at the centre of the disc always has the same value so long as the temperature distribution at the boundary is unchanged. This physical requirement of uniqueness implies a requirement of uniqueness of the solution of the partial differential equation providing a mathematical model of the system. Thus, in this section, we discuss the important question of uniqueness of solutions of the equations treated in this book and, in doing so, it is convenient to consider the equations from the point of view of Dirichlet, Neumann and Cauchy problems defined in Section 4.

For purposes of illustration, at a trivial level, we first consider the simple, second-order, ordinary differential equation,

$$\frac{d^2u}{dx^2} = 0, \tag{1.20}$$

having the general solution, $u = c_1 x + c_2$, where c_1 and c_2 are constants. The solution u is specified uniquely once c_1 and c_2 are determined and we consider two different ways of doing this. First, we suppose u has given values at two distinct values of x, say, at $x = 0$ and $x = l$, and these determine c_1 and c_2 uniquely. Thus we might regard the problem as a Dirichlet problem in the interval, $0 \leqslant x \leqslant l$, for the equation (1.20), the simple one-dimensional form of Laplace's equation, with u prescribed 'on the boundaries' which here consist of just two points.

Alternatively, we could specify the values of u and du/dx at some particular value of x, say $x = 0$. The constants c_1 and c_2 are again uniquely determined but now the problem corresponds to a Cauchy problem and no condition on u can be imposed elsewhere. This initial value problem is akin to that for the wave equation with respect to the time variable. Since $du/dx = $ const. for equation (1.20) it is not possible to prescribe different values of the 'normal derivative', du/dx, at two boundary points such as $x = 0$ and $x = l$.

Guided by the result for the one-dimensional case we now enquire whether the Dirichlet problem for Laplace's equation in two dimensions, equation (1.2c) in Cartesian coordinates, has a unique solution. The answer is given below.

THEOREM 1.1
The solution of the Dirichlet problem in which,

   (a) $\nabla^2 u = \partial^2 u/\partial x^2 + \partial^2 u/\partial y^2 = 0$ inside a region R in the (x, y) plane, bounded by a closed curve C,
   (b) u is specified on C,

is uniquely defined.

We assume the simplest situation (Fig. 1.3) in which R is simply connected and C is a piecewise-smooth curve (that is to say it is continuous and composed of a finite number of curves, or 'pieces', along each of which the unit norm varies continuously in direction).

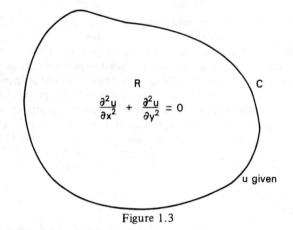

Figure 1.3

Proof.   As in most uniqueness proofs we start by assuming that there are two possible solutions, u_1 and u_2, satisfying all the conditions, and consider their difference $\psi = u_1 - u_2$. Then, from (a) and (b),

$$\nabla^2 \psi = 0 \text{ in } R, \tag{1.21}$$

$$\psi = 0 \text{ on } C. \tag{1.22}$$

Now Green's Theorem states that, for two scalar functions of position, a and b,

$$\int_\tau a\nabla^2 b \, d\tau + \int_\tau (\nabla a . \nabla b) \, d\tau = \int_S a \frac{db}{dn} \, dS,$$

where $\int_\tau . \, d\tau$ represents a volume integral over a volume τ and $\int_S . \, dS$ represents the surface integral over S, which bounds τ. The normal derivative (outward) over S is denoted by d/dn. In the present plane case the integral over τ is a double integral over R and that over S a line integral round C. On setting $a = b = \psi$, we obtain

$$\int_R \psi \nabla^2 \psi \, d\tau + \int_R (\nabla \psi)^2 \, d\tau = \int_C \psi \frac{d\psi}{dn} \, ds,$$

where ds is an element of arc length of C. Hence, taking account of the stated conditions (1.21) and (1.22) on ψ, we deduce that

$$\int_R (\nabla \psi)^2 \, d\tau = \int_R \left\{ \left(\frac{\partial \psi}{\partial x} \right)^2 + \left(\frac{\partial \psi}{\partial y} \right)^2 \right\} d\tau = 0.$$

Since the integrand can only be positive or zero, for the equation to be true we must have

$$\frac{\partial \psi}{\partial x} = 0, \quad \frac{\partial \psi}{\partial y} = 0.$$

Thus $\psi(x, y)$ is independent of x and y and, hence, $\psi =$ const. $= 0$, from (1.22). Since $u_1 = u_2$ the theorem is proved.

In general, of course, $\nabla\psi = 0$ implies ψ is constant so that the choice of coordinate system is irrelevant. Physically, the theorem must obviously hold; for example, the temperature at any point of a metal plate, with insulated faces, and edges maintained at $0°C$ must, in the steady state, be itself $0°C$.

Corresponding theorems can be established for different types of boundary conditions on u or its normal derivative as discussed in the preceding section. Some of these are included in the exercises at the end of the chapter. The theorems are also capable of generalization to a higher number of dimensions.

The uniqueness theorem remains true for plane regions external to a closed curve C, such as the exterior of a circular disc discussed in example (ii) of Section 4.5.2, provided the additional condition that u remains bounded as $\sqrt{(x^2 + y^2)} \to \infty$ is imposed.

Next we consider uniqueness of the wave equation, $\partial^2 u/\partial t^2 = c^2 \partial^2 u/\partial x^2$, with Cauchy conditions applied with respect to the time variable. In order to illustrate the ideas involved we prove the simplest case in which derivatives of u are assumed continuous although this is unnecessarily stringent in applications of the wave equation (see the example in Section 2.4). The result is given in the form of the following theorem.

THEOREM 1.2

The solution of the one-dimensional wave equation,

$$\frac{\partial^2 u}{\partial x^2} = \frac{1}{c^2}\frac{\partial^2 u}{\partial t^2}, \quad 0 < x < l, \, t > 0,$$

with boundary conditions,

$$u(0, t) = f(t), \quad u(l, t) = g(t), \quad t \geqslant 0,$$

and initial conditions,

$$u(x, 0) = r(x), \quad \left[\frac{\partial u(x, t)}{\partial t}\right]_{t=0} = s(x), \quad 0 \leqslant x \leqslant l,$$

where the functions f, g, r and s are assumed to be sufficiently smooth (that is, differentiable as often as required), is uniquely defined.

Proof. As in the previous theorem we suppose there exist two solutions, u_1 and u_2, and consider their difference $\psi = u_1 - u_2$. Then ψ satisfies the wave equation and the homogeneous conditions,

$$\psi(0, t) = \psi(l, t) = 0, \quad t \geqslant 0,$$

$$\psi(x, 0) = \left[\frac{\partial \psi(x, t)}{\partial t}\right]_{t=0} = 0, \quad 0 \leqslant x \leqslant l.$$

Again, we start with an application of Green's Theorem but in this case, in one space dimension, it becomes the usual rule for integration by parts. Thus we consider the integral,

$$\int_0^l \frac{\partial \psi}{\partial t} \frac{\partial^2 \psi}{\partial x^2} \, dx = \left[\frac{\partial \psi}{\partial t} \frac{\partial \psi}{\partial x}\right]_{x=0}^{x=l} - \int_0^l \frac{\partial \psi}{\partial x} \frac{\partial^2 \psi}{\partial t \, \partial x} \, dx.$$

Since ψ vanishes at $x = 0$ and l for all t, so does $\partial \psi / \partial t$ and therefore the first term on the right-hand side vanishes. Hence, on using the equation, we obtain

$$\int_0^l \left\{\frac{1}{c^2} \frac{\partial \psi}{\partial t} \frac{\partial^2 \psi}{\partial t^2} + \frac{\partial \psi}{\partial x} \frac{\partial^2 \psi}{\partial x \, \partial t}\right\} \, dx = 0.$$

Therefore, assuming the integrand is continuous, we may write

$$\frac{1}{2} \int_0^l \frac{\partial}{\partial t} \left\{\frac{1}{c^2}\left(\frac{\partial \psi}{\partial t}\right)^2 + \left(\frac{\partial \psi}{\partial x}\right)^2\right\} \, dx =$$

$$\frac{1}{2} \frac{d}{dt} \int_0^l \left\{\frac{1}{c^2}\left(\frac{\partial \psi}{\partial t}\right)^2 + \left(\frac{\partial \psi}{\partial x}\right)^2\right\} \, dx = 0.$$

On integrating,

$$\int_0^l \left\{\frac{1}{c^2}\left(\frac{\partial \psi}{\partial t}\right)^2 + \left(\frac{\partial \psi}{\partial x}\right)^2\right\} \, dx = \text{const.} = 0,$$

since the integrand vanishes for $0 < x < l$ at $t = 0$. It follows that each term in the integrand, being positive, must separately vanish; $\partial \psi / \partial x = \partial \psi / \partial t = 0$, and so $\psi = \text{const.} = 0$ from its initial value.

Again extensions of the theorem to cover different boundary conditions or weaker differentiability properties can be proved. The proof of uniqueness for the one-dimensional diffusion equation follows along similar lines to that given above. We leave it as an exercise (Exercise 1.6).

Lastly, in this section we mention a further requirement of the solution of our model equations which represent physical phenomena. Since practical measurements of data, which provide initial or boundary conditions, must inevitably contain errors, in most cases we should like the solution to be stable in the sense that small changes in the boundary data should produce only small changes in the solution; in other words the solution must depend continuously on the imposed initial and boundary conditions. For example, it is shown in Chapter 4 that a non-constant solution of Laplace's equation in a bounded domain attains its maximum and minimum values on the boundary. It follows that any small change in boundary values cannot produce bigger changes (in absolute magnitude) in the solution throughout the domain. Also, the general solution of the wave equation, (2.25), derived in Chapter 2 demonstrates explicitly that the solution u depends continuously on the initial values of u and $\partial u/\partial t$.

If a solution of a partial differential equation can be found which is unique and continuously dependent on the boundary data, the problem is said to be properly-posed. All the problems we shall consider have this property.

1.6 Fourier Series

In the following chapters there will be frequent use made of Fourier series as representations of solutions of partial differential equations. In this section we find it convenient to collect together and present, without proof, the main results that we shall need. The principal theorem, establishing existence and convergence of Fourier series, is sometimes known as Dirichlet's theorem and is as follows.

THEOREM 1.3
A single-valued function $f(x')$ which is defined for all x', subject to the conditions that it is periodic with period 2π and in the interval $-\pi < x' < \pi$ has only

 (i) a finite number of discontinuities,
 (ii) a finite number of maxima and minima,

can be uniquely represented by the Fourier series,

$$f(x') = \frac{A_0}{2} + \sum_{n=1}^{\infty} (A_n \cos nx' + B_n \sin nx'),$$

where the constants,

$$A_n = \frac{1}{\pi} \int_{-\pi}^{\pi} f(x') \cos nx' \, dx', \quad n = 0, 1, 2, \ldots,$$

$$B_n = \frac{1}{\pi} \int_{-\pi}^{\pi} f(x') \sin nx' \, dx', \quad n = 1, 2, 3, \ldots . \tag{1.23}$$

The series converges to $f'(x)$ for all x' except that at points of discontinuity it converges to the mean of the values of $f(x')$ on either side of the discontinuity.

The choice of a period of 2π is not restrictive. To modify the theorem so that it applies to periodic functions $f(x)$ with an arbitrary period $2l$

we have only to transform the x' variable by $x' = \pi x / l$. The Fourier series is then

$$f(x) = \frac{A_0}{2} + \sum_{n=1}^{\infty} \left(A_n \cos \frac{n \pi x}{l} + B_n \sin \frac{n \pi x}{l} \right),$$

where

$$\left. \begin{array}{l} A_n = \dfrac{1}{l} \displaystyle\int_{-l}^{l} f(x) \cos \left(\dfrac{n \pi x}{l} \right) dx, \quad n = 0, 1, 2, \ldots, \\[4mm] B_n = \dfrac{1}{l} \displaystyle\int_{-l}^{l} f(x) \sin \left(\dfrac{n \pi x}{l} \right) dx, \quad n = 1, 2, \ldots. \end{array} \right\} \tag{1.24}$$

Of course, we can determine the constants A_n and B_n from (1.24), and hence the Fourier series, for any suitable function defined over a finite interval which, without loss of generality, we may set as $-l < x < l$. Then the sum of such a series for x lying outside this interval is the *periodic extension* of $f(x)$, say $f^*(x)$, which is defined by

$$f^*(x) = f(x), \quad -l < x < l,$$
$$f^*(x + 2l) = f^*(x), \quad \text{all } x.$$

In Fig. 1.4 is drawn the periodic extension of the function, $f(x) = x + l$, $-l < x < l$. The coefficients of the series can be calculated easily from

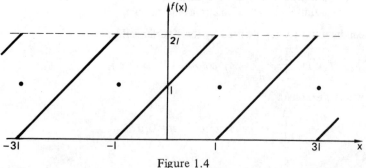

Figure 1.4

the formulae (1.24) and when x is an odd multiple of l the sum of the Fourier series is l, the mid-way point of the discontinuities, as stated in Theorem 1.3.

In many applications we shall be interested in values of x lying in an interval $0 < x < l$. To consider Fourier series of functions defined in this interval we consider two special cases; an extension of $f(x)$, call it

$f_{ex}(x)$, to the full interval, $-l < x < l$, such that $f_{ex}(x)$ is odd or $f_{ex}(x)$ is even. In the former case,

$$f_{ex}(x) = \begin{cases} f(x), & 0 < x < l, \\ -f(-x), & -l < x < 0, \end{cases} \tag{1.25}$$

and in the latter case,

$$f_{ex}(x) = \begin{cases} f(x), & 0 < x < l, \\ f(-x), & -l < x < 0. \end{cases} \tag{1.26}$$

For odd extensions $f_{ex}(0) = 0$ and for even extensions $f_{ex}(0) = \lim_{x \to 0+} f(x)$, where $x \to 0+$ means, 'x approaches zero from the right'.

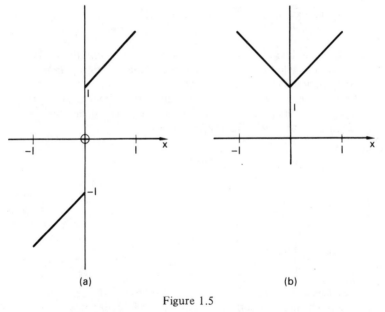

(a)   (b)

Figure 1.5

An illustration is given in Fig. 1.5 for the function $f(x) = x + l$, $0 < x < l$. Notice that neither of these extensions agrees with $f(x) = x + l$ in $-l < x < 0$, drawn in Fig. 1.4.

We can now calculate the Fourier coefficients A_n and B_n for the periodic extensions, $f^*(x)$, of the odd and even functions $f_{ex}(x)$. In the odd case, we obtain

$$lA_n = \int_{-l}^{l} f_{ex}(x) \cos \frac{n\pi x}{l} dx = \left\{ \int_{-l}^{0} + \int_{0}^{l} \right\} \left(f_{ex}(x) \cos \frac{n\pi x}{l} \right) dx.$$

On using the definition (1.25) of $f_{ex}(x)$ and writing $x = -s$ as new variable of integration in the first integral above we deduce that

$$lA_n = \int_l^0 f(s) \cos \frac{n\pi s}{l} \, ds + \int_0^l f(x) \cos \frac{n\pi x}{l} \, dx = 0, \quad n = 0, 1, 2, \ldots.$$

Similarly,

$$B_n = \frac{2}{l} \int_0^l f(x) \sin \frac{n\pi x}{l} \, dx, \quad n = 1, 2, 3, \ldots. \tag{1.27}$$

Thus the constant term and all the coefficients of the cosine terms drop out of the Fourier series. The result, that

$$f(x) = \sum_{n=1}^{\infty} B_n \sin \frac{n\pi x}{l},$$

with B_n given by (1.27), is called the *Fourier sine series* for $f(x)$ in $0 < x < l$. Outside this interval it represents the function $f^*(x)$, which is the odd, periodic extension of $f(x)$.

In the same way we can form the even, periodic extension of $f(x)$ using $f_{ex}(x)$ defined by (1.26). The B_n's then all vanish and we obtain the *Fourier cosine series* for $f(x)$ in $0 < x < l$,

$$f(x) = \frac{A_0}{2} + \sum_{n=1}^{\infty} A_n \cos \frac{n\pi x}{l},$$

for which

$$A_n = \frac{2}{l} \int_0^l f(x) \cos \frac{n\pi x}{l} \, dx.$$

Because, according to Theorem 1.3, a Fourier series can converge to a function which has discontinuities, it may well be that it is not uniformly convergent. Thus term-by-term integrability, which is characteristic of uniformly convergent series seems not, in general, to be assured for Fourier series. However, it is justified, as stated in Theorem 1.4.

THEOREM 1.4
The Fourier series,

$$f(x) = \frac{A_0}{2} + \sum_{n=1}^{\infty} \left(A_n \cos \frac{n\pi x}{l} + B_n \sin \frac{n\pi x}{l} \right),$$

can be integrated term-by-term to give

$$\int_{-l}^{x} f(s) \, ds = \frac{A_0(x+l)}{2} + \sum_{n=1}^{\infty} \frac{l}{n\pi} \left[A_n \sin \frac{n\pi x}{l} - B_n \left(\cos \frac{n\pi x}{l} - \cos n\pi \right) \right],$$

where $-l \leqslant x \leqslant l$.

The result is not in general a Fourier series since it contains a linear term in x. It is noteworthy that the theorem remains valid if $f(x)$ is integrable but other conditions on $f(x)$ are relaxed, so that its Fourier series is not even convergent.

A more important consideration, in view of our intended representation of solutions of partial differential equations as Fourier series, is the question of differentiability. It will be recalled that one condition needed to ensure the justification of term-by-term differentiation of an infinite series is the uniform convergence of the differentiated series. For Fourier series an important requirement is continuity of the periodic function even when the function is defined as the extension of a given function specified over a finite interval, so that the values at its end points must be equal. Theorem 1.5, which follows, gives conditions which guarantee the process of term-by-term differentiation.

THEOREM 1.5
If $f(x)$ is continuous in $-l \leqslant x \leqslant l$, $f(l) = f(-l)$ and $f' = df/dx$ is piecewise continuous, then the Fourier series,

$$f(x) = \frac{A_0}{2} + \sum_{n=1}^{\infty} \left(A_n \cos \frac{n\pi x}{l} + B_n \sin \frac{n\pi x}{l} \right), \quad -l \leqslant x \leqslant l,$$

can be differentiated term-by-term wherever f' is itself differentiable; at such points,

$$f'(x) = \sum_{n=1}^{\infty} \frac{n\pi}{l} \left(-A_n \sin \frac{n\pi x}{l} + B_n \cos \frac{n\pi x}{l} \right), \quad -l < x < l.$$

EXERCISES
1. Verify that if u_1 is a solution of $Lu = f$ (equation (1.3)), where L is a linear partial differential operator, and u_2 a solution of $Lu = 0$, then $u_1 + cu_2$ also satisfies $Lu = f$ for any constant c.

2. Consider the simple nonlinear ordinary differential equation, $dy/dx + y^n = 0$ for $n = 2, 3, 4, \ldots$. Show that $y(x; k) = [(n-1)x + k]^{-1/(n-1)}$, k arbitrary, is the general solution but that the sum of two solutions, $y(x; k_1) + y(x; k_2)$, $(k_1 \neq k_2)$, does not satisfy the equation. Discuss the linear case $n = 1$.

3.   Verify that the functions $u_1 = \cos[n(t - x/c)]$ and $u_2 = \cos[n(t + x/c)]$ satisfy the wave equation, $\partial^2 u/\partial t^2 - c^2\, \partial^2 u/\partial x^2 = 0$ (cf. Chapter 2). Deduce, using superposition, that a solution which vanishes at $x = 0$ for all t is $\sin nt\, \sin(nx/c)$.

4.   Determine in which regions of the (x, y) plane the equation,

$$ax\,\frac{\partial^2 u}{\partial x^2} + 2hy\,\frac{\partial^2 u}{\partial x\,\partial y} + bx\,\frac{\partial^2 u}{\partial y^2} = 0, \quad a, b, h \text{ constant,}$$

is elliptic, parabolic and hyperbolic.

5.   Consider continuous solutions of the wave equation, with continuous first derivatives, in two regions of the (x, t) plane separated by a curve C: $x = x(t)$. Across C, $\partial^2 u/\partial x^2$, $\partial^2 u/\partial t\,\partial x$, $\partial^2 u/\partial t^2$ are discontinuous but along C derivatives of $\partial u/\partial x$ and $\partial u/\partial t$ are continuous. By using a formula similar to (1.16) for differentiation along C and considering the jumps $[\partial^2 u/\partial t^2]$, $[\partial^2 u/\partial x\,\partial t]$ and $[\partial^2 u/\partial x^2]$ in the second derivatives across C, deduce that C must be given by $dx/dt = \pm c$. Show that C must be a characteristic for the wave equation. (Use (1.18).)

6.   Prove uniqueness of solutions of the initial value problem for the diffusion equation (1.2b) in $0 < x < l$ satisfying the boundary conditions $u(0, t) = f(t)$, $u(l, t) = g(t)$, $t \geqslant 0$. (Hint. The difference of two solutions, ψ, having the same initial value in $0 < x < l$ satisfies (1.2b) with zero boundary and initial conditions. Multiply this by ψ and integrate over the interval $0 < x < l$.)

7.   Detect at what point the proof of Theorem 1.2 is altered when boundary conditions on the wave equation are given in the form of pre-scribed values of $\partial u/\partial x$, instead of u, at either or both points $x = 0$ and $x = l$.

8.   If, in Theorem 1.1, the normal derivative du/dn, instead of u, is given on C, prove that the solution of Laplace's equation is unique to within an arbitrary constant. (Neumann problem.)

9.   If, in Theorem 1.1, the quantity $du/dn + hu$, where h is a positive constant, is given on the boundary, prove that the solution of Laplace's equation is unique to within an arbitrary constant.

10.   Demonstrate that it is not possible to formulate the Neumann problem for the hyperbolic equation $\partial^2 u/\partial x\,\partial y = 0$ in the rectangle, $0 \leqslant x \leqslant a,\ 0 \leqslant y \leqslant b$ (a, b constant).

11.   (a) Find the Fourier series for the function $f(x) = 0$, $-l < x < 0$; $f(x) = l - x$, $0 < x < l$.
   (b) Find the Fourier sine series for the function $f(x) = l - x$, $0 < x < l$.

(c) Find the Fourier cosine series for the function in part (b). Verify that the series obtained in parts (b) and (c) add up to twice the series obtained in part (a) in the interval $-l < x < l$. What is the sum of each series at the point $x = 0$?

12.   Find the Fourier cosine series in $0 < x < l$ for the function $f(x) = \sin(\pi x/l)$. To what function does the series converge in $-l < x < 0$?

13.   In some applications (see Example (ii) of Section 2.7) it is required to find the coefficients C_n in a series of the type,

$$f(x) = \sum_{n=0}^{\infty} C_n \cos \frac{(2n + 1)\pi x}{2l} \, ,$$

where $f(x)$ is a given function in $0 < x < l$. By considering the Fourier cosine series for $f^*(x) = f(x), 0 < x < l; f^*(x) = -f(2l - x), l < x < 2l$; over the double range $0 < x < 2l$, show that

$$C_n = \frac{2}{l} \int_{0}^{l} f(x) \cos \left[\frac{(2n + 1)\pi x}{2l} \right] dx, \quad n = 0, 1, 2, \ldots$$

CHAPTER 2

The Wave Equation

2.1 Introduction

This chapter is concerned with the study of the partial differential equation of second order,

$$\frac{\partial^2 u}{\partial t^2} = c^2 \frac{\partial^2 u}{\partial x^2},$$ (2.1)

called the one-dimensional wave equation. The independent variables x and t denote the space coordinate and time respectively. Depending on the physical situation under study, the dependent variable u has many possible interpretations. Inspection of equation (2.1) shows that the constant c has dimensions of velocity; this speed represents the rate at which a wave travels in the x (or $-x$) direction (see Section 2.3). Thus the wave equation can transmit information only at a finite speed that is characteristic of the medium.

Equation (2.1) is the prototype for second-order hyperbolic partial differential equations and it exhibits many important properties typical of them. It arises in a great variety of physical situations some of which are described in Section 2.2. As it and its initial and boundary conditions are linear we make extensive use of the Principle of Superposition (Section 1.2) to construct solutions of particular problems in Sections 2.3, 2.5 and 2.7. Sections 2.3–2.5 are mainly concerned with the solution of (2.1) in an infinite medium, $-\infty < x < \infty$, and (2.7) in a finite medium, say $0 < x < l$, for which a proper choice of boundary conditions at $x = 0$ and l must be made as described in Section 2.6. In Section 2.8 the behaviour of the energy in wave motion is discussed.

2.2 Applications of the Wave Equation

2.2.1 Transverse Waves on Strings

Everyone has jerked one end of a rope up and down and seen how a 'hump' or sequence of 'humps' of rope appear to move along its length. This is a crude example of a transverse wave propagating along the rope. The adjective 'transverse' describes the manner in which particles of the rope move at right angles to the direction in which the wave moves. This is only roughly so in the crude example given above but in the idealized mathematical model that we adopt for a stretched elastic string it is assumed to hold exactly. Thus the components of acceleration and

24

velocity of the string particles parallel to the undisturbed (straight-line) position of the string are negligible compared with those normal to this direction.

To be definite, we suppose that in equilibrium the mass per unit length ρ_0 and tension T_0 are constant, with T_0 so large that effects of gravity can be ignored compared with it. Since, by Hooke's Law, tension is proportional to extension, the string must be stretched beyond its natural length, and we assume it has its ends fixed at two points along the x-axis. When disturbed (for example, by plucking) the string departs from equilibrium and acquires velocity and acceleration. If the motion occurs in the (x, z) plane, the situation is as illustrated in Fig. 2.1. Here,

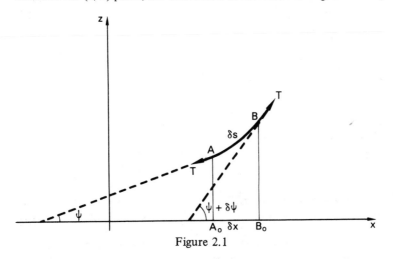

Figure 2.1

at a given instant, an element of the string AB, of length δs, is drawn to represent those particles of the string which, in the equilibrium position, form the projection $A_0 B_0$, of length δx, of this element onto the x-axis. Since the string is flexible the tension at any point acts along the tangent which makes an angle ψ with the x-axis. The x and z components of the force on AB due to the tension are, respectively,

$$(T \cos \psi)_{\psi + \delta \psi} - (T \cos \psi)_\psi, \quad (T \sin \psi)_{\psi + \delta \psi} - (T \sin \psi)_\psi.$$

$$(2.2a, b)$$

The x-component vanishes if the element moves always parallel to the z-axis and hence for small $\delta \psi$, $\partial/\partial \psi \, (T \cos \psi)\delta \psi = 0$. Thus, $T \cos \psi =$ const. along the string.

Now we make the further assumption that ψ remains small for all x and t. Then $\cos \psi = 1 + O(\psi^2)$, where $O(\psi)$ means the order of magnitude is 4 as ψ approaches zero and hence T can be taken to be uniform and equal to T_0 because the error in so doing is proportional to $\delta s - \delta x = \delta x (\sec \psi - 1) = \delta x O(\psi^2)$.

The z-component of the equation of motion of AB, whose mass is $\rho_0\,\delta x$, is

$$\rho_0\,\delta x\,\frac{\partial^2 \bar{z}}{\partial t^2} = T_0 \cos\psi\,\delta\psi + 0(\delta\psi)^2,$$

where $\partial^2\bar{z}/\partial t^2$ is the acceleration of its mass-centre and the force comes from (2.2b). On dividing by δx and letting δx tend to zero the last term vanishes if $\partial\psi/\partial x$ is bounded (so that the slope of the string remains finite) and we obtain, in our small ψ approximation,

$$\rho_0\,\frac{\partial^2 z}{\partial t^2} = T_0\,\frac{\partial\psi}{\partial x}. \tag{2.3}$$

The variable ψ can be eliminated from (2.3) by using $\partial z/\partial x = \tan\psi$ to give

$$\frac{\partial^2 z}{\partial t^2} = c^2\,\frac{\partial^2 z}{\partial x^2}, \quad c^2 = \frac{T_0}{\rho_0}, \tag{2.4}$$

the wave equation for the string.

The motion of violin and piano strings may be taken as practical examples of these wave motions and is discussed further in later sections.

2.2.2 Sound Waves in a Gas

The sense of hearing is the detection of pressure fluctuations in the atmosphere by the ear. These pressure changes, produced by some external source such as a person speaking, are what we call sound. Hence in examining how sound propagates it is essential to have some medium to act as the propagating agent; no sound is propagated in a vacuum. We choose the medium to be a gas as this is of everyday importance, but we could equally well choose a solid or a liquid.

Here, rather than proceed directly from first principles, we start with the appropriate general partial differential equations governing the behaviour of the pressure (p), density (ρ) and velocity (v) of a gas and simplify them for the conditions relevant to sound propagation. Then in one-dimensional motion, the equations expressing the conservation of mass of an arbitrary volume of gas as it moves, and that equating the rate of change of its momentum to the pressure forces acting on it are, respectively,

$$\frac{\partial\rho}{\partial t} + \frac{\partial}{\partial x}(\rho v) = 0, \tag{2.5}$$

$$\rho\,\frac{\partial v}{\partial t} + \frac{\rho}{2}\,\frac{\partial}{\partial x}(v^2) = -\frac{\partial p}{\partial x}. \tag{2.6}$$

In addition we require the adiabatic law,

$$p = K\rho^{\gamma}, \quad K, \gamma \text{ consts.,} \tag{2.7}$$

which corresponds to the physical assumption that effects of viscosity and heat conduction in the gas are negligible for the motion under consideration.

Again, we assume a uniform equilibrium state (with the gas at rest) which is slightly disturbed; for example, by blowing across the end of an open cylindrical tube to produce a note. We characterize the disturbance by dimensionless barred variables, defined by the equations,

$$v = v_0\bar{v}, \quad \rho = \rho_0(1 + \bar{\rho}), \quad p = p_0(1 + \bar{p}), \tag{2.8}$$

where $\bar{v}, \bar{\rho}, \bar{p} \ll 1$ (all x, t), p_0 and ρ_0 are equilibrium values and v_0 is a constant velocity to be specified later. These expressions are substituted into equations (2.5) and (2.6) and all terms involving the square or product of barred variables are neglected. The approximate form of equation (2.7) is obtained by expanding $(1 + \bar{\rho})^{\gamma}$ as far as the linear term and using the fact that (2.7) itself is satisfied by the equilibrium values p_0 and ρ_0. The resulting equations are

$$v_0 \frac{\partial \bar{v}}{\partial t} + \frac{p_0}{\rho_0} \frac{\partial \bar{p}}{\partial x} = 0, \tag{2.9}$$

$$\frac{\partial \bar{\rho}}{\partial t} + v_0 \frac{\partial \bar{v}}{\partial x} = 0, \tag{2.10}$$

$$\bar{p} = \gamma \bar{\rho}. \tag{2.11}$$

This process of linearizing the equations is widely used to simplify the mathematical model. The variable $\bar{p}$ can be eliminated by using (2.11) and then by simple partial differentiation $\bar{v}$ and $\bar{\rho}$ can be eliminated in turn from the remaining equations. We find that each of the variables $\bar{v}$, $\bar{\rho}$ (and hence $\bar{p}$) satisfies the wave equation,

$$\frac{\partial^2 u}{\partial t^2} = c^2 \frac{\partial^2 u}{\partial x^2}, \quad c^2 = \frac{\gamma p_0}{\rho_0}. \tag{2.12}$$

Thus the sound speed c arises naturally in the problem and the arbitrary speed v_0 may be set equal to it. It follows that the magnitude of the second term in (2.6) is small compared with the others if the fluid velocity changes induced by the pressure fluctuations are much smaller than the sound speed.

In this case the gas particles move always in the x-direction which is the direction of propagation of the wave. Waves having this property are called *longitudinal*.

2.2.3 Electromagnetic Waves in a Cable

Next we consider the flow of electricity in a linear conductor such as a telephone wire or submarine cable. The circuit in Fig. 2.2 consists of a current source S, a long cable AB whose length at any point is denoted by x, a load W at the receiving end and a return via earth. The current in the cable is $I(x, t)$, the e.m.f. $V(x, t)$, and the capacitance to earth and inductance, C and L per unit length, respectively. Provided we restrict the discussion to high frequencies we may regard the cable as 'lossless'; that is, the resistance of the cable and the conductance to earth through the insulating sheath can be neglected.

Hence across an element of cable δx, the changes in potential δV and current δI are entirely due to the cable's inductance and capacitance, respectively, and are of amount,

$$\delta V = -\frac{\partial I}{\partial t} L\, \delta x, \quad \delta I = -\frac{\partial V}{\partial t} C\, \delta x. \tag{2.13}$$

By dividing by δx and letting δx tend to zero we obtain the equations,

$$\frac{\partial V}{\partial x} = -L\frac{\partial I}{\partial t}, \quad \frac{\partial I}{\partial x} = -C\frac{\partial V}{\partial t}.$$

On eliminating I and V in turn from these equations we see that V and I each satisfies the wave equation,

$$\frac{\partial^2 u}{\partial t^2} = c^2 \frac{\partial^2 u}{\partial x^2}, \quad c^2 = (LC)^{-1}$$

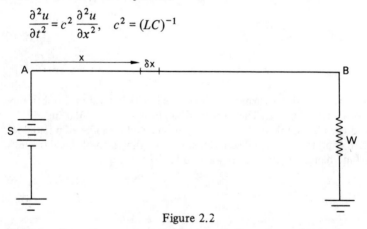

Figure 2.2

2.2.4 Water Waves in a Channel

A familiar situation in which waves may be observed is on the surface of water. To make the illustration as simple as possible, we suppose the water is contained within a long, straight, horizontal channel of rectangular cross-section. We choose the coordinate axes as in Fig. 2.3, where the x-axis lies along the undisturbed water level. The mean depth of the water is h, so that the bottom of the channel is at $z = -h$. We assume that the water possesses only one component of velocity v, parallel to the x-axis, and this means that the pressure p at level z is determined by the hydrostatic law. If the water level at the point x and time t is $z = Z(x, t)$,

$$p = p_0 + \rho g(Z - z), \tag{2.14}$$

where ρ is the constant density of the water, g the acceleration due to gravity and p_0 is the atmospheric pressure.

The equation of motion for the fluid is again (2.6). For sufficiently small velocities the nonlinear terms in (2.6) may be neglected and using equation (2.14) we obtain

$$\frac{\partial v}{\partial t} = -g\,\frac{\partial Z}{\partial x}. \tag{2.15}$$

In contrast to Section 2.2.2, the mass of fluid in an elementary volume of length δx can vary because the volume changes (through the varying height Z) rather than because the density changes. Thus the mass M of liquid, flowing in time δt across the face AB (Fig. 2.3), is approximately that within a length $v\,\delta t$ to the left of AB, or

$$M = \rho(v\,\delta t)b(h + Z),$$

where b is the channel width. By Taylor's expansion the outflow through CD in the same time has mass $M + \partial M/\partial x\,\delta x$, approximately. Hence the

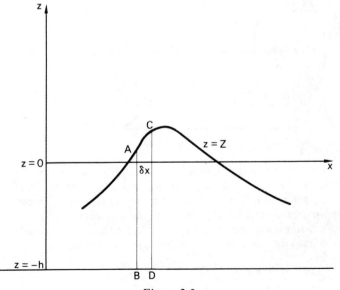

Figure 2.3

net mass inflow to $ABCD$ in time δt is approximately

$$-\rho bh \frac{\partial v}{\partial x} \delta x \, \delta t, \tag{2.16}$$

if we assume $Z \ll h$. The minus sign indicates an outflow.

The approximate mass of liquid in $ABCD$, $\rho (h + Z)b \, \delta x$, increases by $\rho b (\partial Z/\partial t) \, \delta x \, \delta t$ in time δt and this must balance the inflow (2.16) since no fluid is created or destroyed. Hence, $h \, \partial v/\partial x = -\partial Z/\partial t$, which combined with (2.15) yields the wave equation,

$$\frac{\partial^2 u}{\partial t^2} = c^2 \frac{\partial^2 u}{\partial x^2}, \quad c^2 = gh, \tag{2.17}$$

for either v or Z.

2.2.5 Torsional Waves on a Rod

Consider a uniform straight rod of circular cross-section whose axis is the x-axis. When the rod is twisted so that the angle of twist of a cross-section at a station x along its length is ϕ relative to that at the origin the torque exerted by the material to the left of x on that to the right is $-K \, \partial \phi/\partial x$, where the constant K depends on the elastic properties and dimensions of the rod.

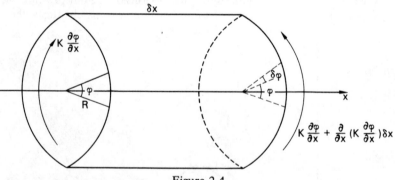

Figure 2.4

For the slice of rod of length δx shown in Fig. 2.4 the moment of inertia about the x-axis is $\rho \, \delta x R^2/2$, where ρ is the mass per unit length and R is the radius. Hence the rate of change of its angular momentum is approximately $(\rho R^2/2)(\partial^2 \phi/\partial t^2) \, \delta x$, and this must equal the net torque acting on the element. From the figure this is approximately $K(\partial^2 \phi/\partial x^2) \, \delta x$, and therefore ϕ satisfies the wave equation,

$$\frac{\partial^2 \phi}{\partial t^2} = c^2 \frac{\partial^2 \phi}{\partial x^2}, \quad c^2 = \frac{K}{I},$$

where I is the moment of inertia about the x-axis per unit length of rod. The particles of the material move along arcs of circles in planes normal to the direction of wave propagation and therefore the waves are transverse.

In Section 2 we have indicated a few of the situations in which the one-dimensional wave equation arises. In each case the equation has been obtained by making simplifying assumptions about the types of motion being considered. The basic and crucial one which is common to most situations is the restriction on the magnitude of the motion in some sense (that is, on the size of the dependent variables and their derivatives) as it is this which enables us to arrive at a linear equation for the motion. The illustrations described include motions of gases, liquids and solids as well as the behaviour of current in a conductor, and these examples could be multiplied many times. Thus even the simplest form of wave equation that we are concerned with in this book can be seen to have a wide range of applications in the physical world.

2.3 D'Alembert's Solution of the Wave Equation

A standard method for attempting the solution of both partial and ordinary differential equations is to change variables in some way. There is no prior guarantee that a solution will be obtained thereby (indeed for the vast majority of equations it is impossible to write down a mathematical function for the solution by any method) but in some cases a change of variable can be very fruitful. The wave equation provides an example. Although there is no definite rule for deciding on the particular choice of new variables that will prove most advantageous, past experience can be of assistance. Occasionally, however, as in the present example, the equation is so simple in form that the transformation to new independent variables can be left arbitrary to begin with and the specific choice to ensure maximum simplification can then be made when the equation is obtained in the new variables. Thus we introduce new variables, ξ and η, related to x and t by the equations,

$$\xi = F(x, t), \quad \eta = G(x, t), \tag{2.18}$$

where we assume that the otherwise arbitrary functions F and G are differentiable to as many orders as needed and that the transformation of variables (2.18) is non-singular so that the Jacobian $\partial(\xi, \eta)/\partial(x, t) = \partial F/\partial x \, \partial G/\partial t - \partial F/\partial t \, \partial G/\partial x \neq 0$. Upon applying the rules for partial differentiation we obtain,

$$\frac{\partial}{\partial x} = \frac{\partial F}{\partial x} \frac{\partial}{\partial \xi} + \frac{\partial G}{\partial x} \frac{\partial}{\partial \eta}, \quad \frac{\partial}{\partial t} = \frac{\partial F}{\partial t} \frac{\partial}{\partial \xi} + \frac{\partial G}{\partial t} \frac{\partial}{\partial \eta},$$

$$\frac{\partial^2}{\partial x^2} = \frac{\partial^2 F}{\partial x^2}\frac{\partial}{\partial \xi} + \frac{\partial^2 G}{\partial x^2}\frac{\partial}{\partial \eta} + \left(\frac{\partial F}{\partial x}\right)^2 \frac{\partial^2}{\partial \xi^2} + 2 \frac{\partial F}{\partial x}\frac{\partial G}{\partial x}\frac{\partial^2}{\partial \xi \partial \eta} +$$
$$+ \left(\frac{\partial G}{\partial x}\right)^2 \frac{\partial^2}{\partial \eta^2},$$

$$\frac{\partial}{\partial t^2} = \frac{\partial^2 F}{\partial t^2}\frac{\partial}{\partial \xi} + \frac{\partial^2 G}{\partial t^2}\frac{\partial}{\partial \eta} + \left(\frac{\partial F}{\partial t}\right)^2 \frac{\partial^2}{\partial \xi^2} + 2 \frac{\partial F}{\partial t}\frac{\partial G}{\partial t}\frac{\partial^2}{\partial \xi \partial \eta} +$$
$$+ \left(\frac{\partial G}{\partial t}\right)^2 \frac{\partial^2}{\partial \eta^2}.$$

The wave equation (2.1) is transformed into

$$\frac{\partial^2 u}{\partial \xi^2}\left[c^2\left(\frac{\partial F}{\partial x}\right)^2 - \left(\frac{\partial F}{\partial t}\right)^2\right] + 2\frac{\partial^2 u}{\partial \xi \partial \eta}\left(c^2 \frac{\partial F}{\partial x}\frac{\partial G}{\partial x} - \frac{\partial F}{\partial t}\frac{\partial G}{\partial t}\right) +$$
$$+ \frac{\partial^2 u}{\partial \eta^2}\left[c^2\left(\frac{\partial G}{\partial x}\right)^2 - \left(\frac{\partial G}{\partial t}\right)^2\right] + \frac{\partial u}{\partial \xi}\left(c^2 \frac{\partial^2 F}{\partial x^2} - \frac{\partial^2 F}{\partial t^2}\right) +$$
$$+ \frac{\partial u}{\partial \eta}\left(c^2 \frac{\partial^2 G}{\partial x^2} - \frac{\partial^2 G}{\partial t^2}\right) = 0. \tag{2.19}$$

We should now like to simplify this equation by choosing F and G in some suitable way. The question of simplification is itself somewhat arbitrary but with the aim of eliminating two terms involving second derivatives of u whilst still retaining what symmetry there is in the equation we choose

$$\frac{\partial F}{\partial t} = -c \frac{\partial F}{\partial x}, \quad \frac{\partial G}{\partial t} = +c \frac{\partial G}{\partial x}. \tag{2.20}$$

The minus sign is needed in one of the equations (2.20) because otherwise the transformation (2.18) would be singular since its Jacobian would vanish. The first of equations (2.20) yields

$$\frac{\partial F/\partial t}{\partial F/\partial x} = -\frac{dx}{dt} = -c.$$

Thus, whatever function we choose for F, the independent variables x and t must be combined together in the form $x - ct$. The simplest choice to make is the linear function, $\xi = F(x, t) = x - ct$. The same argument applies to G and there we set, $\eta = G(x, t) = x + ct$. With F and G chosen in this way the differential equation (2.1) becomes

$$\frac{\partial^2 u}{\partial \xi \partial \eta} = 0.$$

From the mathematical viewpoint this is indeed a simpler version of the wave equation because it can be integrated immediately with respect to η to give $\partial u/\partial \xi = f_1(\xi)$ and again with respect to ψ to give $u = f(\xi) + g(\eta)$, where f_1, f and g are arbitrary functions. Expressed in terms of the original variables this is

$$u = f(x - ct) + g(x + ct), \qquad (2.21)$$

and is known as *D'Alembert's general solution* of the wave equation. The arbitrary functions f and g are determined from the way in which the motion is set up as explained in the next section.

Before discussing the properties of the solution (2.21) we may enquire about the significance of the new variables $\xi = x - ct$, $\eta = x + ct$. We first introduce the Cartesian (x, t) plane, drawn in Fig. 2.5, each point of which represents the position of a point x (whether it be on a stretched string, in a moving gas, on the surface of a liquid, etc.) at time t. It says nothing about the value of u at this point (x, t). For that purpose we first

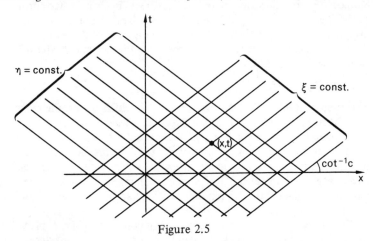

Figure 2.5

observe that constant values of ξ and η represent families of parallel, straight lines in the (x, t) plane making angles of $\cot^{-1}c$ and $\pi - \cot^{-1}c$ respectively with the x-axis. If we consider for the moment the simpler solution in which $g = 0$ in (2.21) and therefore $u = f(x - ct)$ we see that along any one of the straight lines $\xi = $ const. in the (x, t) plane u has a constant value. Thus a knowledge of u at some point P in the (x, t) plane carries with it a knowledge of u at all points along the straight line through P of slope $1/c$. In the system, therefore, as time increases, the value of u at a point corresponding to P is 'passed on' to other points of the system in such a way that $x - ct$ remains constant. The rate at which such points are 'reached' by this value of u is determined by $dx/dt = c$. The speed c (> 0) is the speed of propagation of values of u, that is the wave speed, in the direction of increasing x ('to the right').

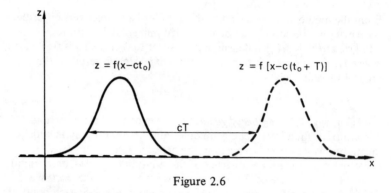

Figure 2.6

The *wave profile* is the function $u(x, t_0) = f(x - ct_0)$ at some instant of time t_0 and is easiest to visualize for transverse waves such as occur on the stretched string. It is there just the shape of the string at $t = t_0$. Suppose that on the string there is a right-travelling wave whose profile at $t = t_0$ is as given in Fig. 2.6 and the wave (the range of x for which the displacement $z \neq 0$) is then far from the end-points of the string. Then the profile moves to the right with speed c so that at time T later it has the equation (Fig. 2.6),

$$z = f[x - c(t_0 + T)].$$

This type of solution of the wave equation in which the profile moves in one direction or the other is called a *travelling wave solution*.

A further important property of the motion, that emerges from this solution, is that there is no change of shape of the profile as the wave moves. The wave is said to be free from distortion. Effects which lead to a change of shape of the profile, such as damping or nonlinearity, have been neglected in deriving the model equation for one-dimensional rectilinear wave motion.

For the wave equation the lines $x - ct = $ const. in the (x, t) plane are one family of characteristics, defined in Section 1.3. By considering the solution $u = g(x + ct)$ the second family is given by $x + ct = $ const. implying that values of u are propagated with velocity $-c$ (to the left). In general, where the full solution (2.21) applies, there is propagation both to the right and left with speed c.

The meaning of the characteristics for the wave equation can be clarified by 'factorizing' the operator as

$$\frac{\partial^2 u}{\partial t^2} - c^2 \frac{\partial^2 u}{\partial x^2} = \left(\frac{\partial}{\partial t} - c \frac{\partial}{\partial x}\right)\left(\frac{\partial u}{\partial t} + c \frac{\partial u}{\partial x}\right) = \left(\frac{\partial}{\partial t} - c \frac{\partial}{\partial x}\right)\left(\frac{du}{dt}\right) = 0,$$

$$(2.22)$$

where du/dt is to be calculated along curves for which $dx/dt = c$ (cf. (1.16)). Thus a function u which is constant along lines $x - ct = $ const.

satisfies (2.1) and, similarly, so does a function constant along lines $x + ct = $ const. The method of finding characteristics described at the end of Section 1.3 can be applied to the wave equation after first writing down the two equivalent first-order equations. It is left as an exercise to verify that D'Alembert's solution results.

2.4 Initial Conditions and Determination of the Solution

The functions f and g which appear in D'Alembert's solution of the wave equation (2.21) are arbitrary. In order to determine them we must take into account the manner in which the wave motion is set up, say at time $t = 0$. We suppose also that the medium is infinite in extent. In practice this means that the splitting up of an initial local disturbance into a simple right and left travelling wave of unchanging form according to (2.21) is valid only so long as no boundaries of the medium are encountered. Across such boundaries, as discussed in Section 2.5, the waves cannot proceed unimpeded.

The equation is of second order in t and therefore two conditions are required on the dependent variable at fixed t for all x. These are usually taken to consist of a knowledge of the initial wave profile and its rate of change, say,

$$u(x, t) = r(x), \quad \frac{\partial u(x, t)}{\partial t} = s(x), \quad \text{at } t = 0, \tag{2.23}$$

where $r(x)$ and $s(x)$ are known functions of x defined for $-\infty < x < \infty$. In the case of transverse waves on a stretched string, for example, these conditions correspond to the initial distribution of displacement and velocity along the string.

In order that $u(x, t) = f(x - ct) + g(x + ct)$ should satisfy these conditions we must have,

$$f(x) + g(x) = r(x),$$

$$\left\{ -c \frac{df(\xi)}{d\xi} + c \frac{dg(\eta)}{d\eta} \right\}_{t=0} = -c \frac{df(x)}{dx} + c \frac{dg(x)}{dx} = s(x),$$

$$\tag{2.24a, b}$$

for all x, where $\xi = x - ct$, $\eta = x + ct$. This is a pair of functional equations for $f(x)$ and $g(x)$. The second can be integrated to give

$$f(x) - g(x) = -\frac{1}{c} \int_{K}^{x} s(\sigma)\, d\sigma,$$

where K is an arbitrary constant, and this result can be combined with the first equation to yield

$$f(x) = \frac{1}{2}\left[r(x) - \frac{1}{c}\int_K^x s(\sigma)\,d\sigma\right], \quad g(x) = \frac{1}{2}\left[r(x) + \frac{1}{c}\int_K^x s(\sigma)\,d\sigma\right],$$

for all x. According to the prescription (2.21) for the solution $u(x, t)$, we deduce that

$$u(x, t) = \frac{1}{2}\left[r(x - ct) + r(x + ct) + \frac{1}{c}\int_{x-ct}^{x+ct} s(\sigma)\,d\sigma\right]. \qquad (2.25)$$

This solution is also known as D'Alembert's solution of the wave equation. The problem just discussed is an example of an initial value problem described in Section 1.4.

Example.   As an example we discuss 'weak' wave propagation in a shock tube, an experimental device used in fluid dynamics for studying high speed flow phenomena in gases. It consists of a long tube which is initially separated into two chambers by a diaphragm at $x = 0$. The state of the gas (pressure, temperature, etc.), initially at rest, is different in the two chambers. At $t = 0$ the diaphragm is broken and the resulting flow of gas constitutes the motion.

If the pressure difference across the diaphragm is sufficiently small the disturbance is said to be weak and the motion is governed approximately by the one-dimensional wave equation, say for the dimensionless pressure perturbation as described in Section 2.2.2. We may choose $\bar{p} = 0$ initially in the low pressure chamber and hence one initial condition is

$$\bar{p} = \begin{cases} 0, & x < 0, \\ \text{const.} = P, & x > 0. \end{cases} \qquad (2.26)$$

To obtain the initial condition on $\partial\bar{p}/\partial t$ we recall from (2.11) and (2.10) that it is proportional to $\partial\bar{v}/\partial x$. Since initially $\bar{v} = 0$ for all x it follows that we must also satisfy

$$\partial\bar{p}/\partial t = 0, \quad \text{all } x, \quad t = 0. \qquad (2.27)$$

The fundamental solution, $\bar{p} = f(x - ct) + g(x + ct)$, must be chosen to satisfy these conditions. From the first condition (2.26), $f(x) + g(x) = 0$ for $x < 0$, and P for $x > 0$, and from (2.27), $f(x) - g(x) = K$, for all x, where K is an arbitrary constant. The solution of these equations for f and g is $f = -g = K/2$ if $x < 0$ and $f = (P + K)/2$, $g = (P - K)/2$, if $x > 0$. Hence

$$\bar{p} = \begin{cases} 0, & x < -ct, \\ P/2, & -ct < x < ct, \\ P, & x > ct. \end{cases}$$

The original pressure discontinuity of amount P splits into two discontinuities (or waves) of amount $P/2$ which propagate with speed c in opposite directions. A pressure profile is drawn in Fig. 2.7 at time t.

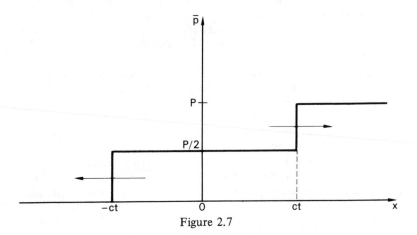

Figure 2.7

Notice that the discontinuities in $\bar{p}$ in the solution occur at points given by $x = \pm ct$, that is along characteristics in the (x, t) plane. This is an example of the property of hyperbolic equations, mentioned in Section 1.3, that any discontinuities in the second derivatives occur across characteristics; the fact that in this case the solution itself is discontinuous is a result of the discontinuity in the initial conditions.

Other physically important variables can be found from the solution for $\bar{p}$. For example, equations (2.9)–(2.11), with v_0 replaced by c, may be written as

$$\gamma \frac{\partial \bar{v}}{\partial t} = -c \frac{\partial \bar{p}}{\partial x}, \quad \gamma \frac{\partial \bar{v}}{\partial x} = -\frac{1}{c} \frac{\partial \bar{p}}{\partial t}.$$

For $\bar{p} = f(x - ct) + g(x + ct)$, these equations can easily be solved for $\bar{v}$ to give $\gamma \bar{v} = f(x - ct) - g(x + ct) - K$, where the constant $-K$ is needed to satisfy the condition that $\bar{v} = 0$ for all x at $t = 0$. Hence,

$$\bar{v} = \begin{cases} 0 & x < -ct, \\ -P/2\gamma, & -ct < x < ct, \\ 0, & x > ct. \end{cases}$$

The effect of the wave motion is to cause the gas particles to move in the negative x-direction between the wave fronts, as we should expect physically. For a tube of finite length the above solution would hold for the whole tube for times $t < T$ where T is the time that a wave front first reaches an end of the tube.

We next indicate the relation between the initial value problem discussed above and the characteristics defined in Section 2.3. Suppose the
initial values of u and $\partial u/\partial t$ are known only along some finite range of
x, say for $a < x < b$ in Fig. 2.8. The object is to find out for what range
of values of x and t, or region of the (x, t) plane, we are able to determine the solution uniquely. Thus the initial conditions are those given by
equation (2.23) but now $r(x)$ and $s(x)$ are known only for $a < x < b$. By
the same sequence of steps which led to equation (2.25) we obtain the
same expressions for $f(x)$ and $g(x)$ but they are valid only for $a < x < b$
since r and s are undefined elsewhere. It follows that the solution (2.25)
is defined only in the region of the (x, t) plane for which both $a < x -$
$ct < b, a < x + ct < b$. The region $a < x - ct < b$, drawn in Fig. 2.8, is
the interior of an inifinite strip in the (x, t) plane with slope $1/c$ and
bounded by the corresponding characteristics through $(a, 0)$ and $(b, 0)$.
Similarly, $a < x + ct < b$ is a strip with slope $-1/c$ bounded by the
corresponding characteristics through the same points. The solution is
therefore determined uniquely within the intersection ABC in Fig. 2.8

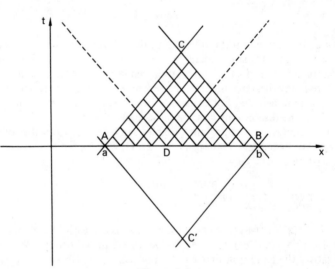

Figure 2.8

of these two strips. This region is called the *domain of dependence* of C
because the solution at C depends only on conditions along the initial
segment $a < x < b$ and is uninfluenced by initial values outside AB. Conversely we can show that conditions at a point D, say, influence the solution only within the region bounded by the characteristics through D.
The latter is referred to as the *region of influence* of D.

Incidentally we note from Fig. 2.8 that a knowledge of u and $\partial u/\partial t$
along AB determines the solution also within the region ABC' which lies

within the half-plane $t < 0$. From a physical point of view this would be appropriate if the scale of t were chosen so that $t = 0$ coincided with the instant at which a camera 'froze' the motion. The solution in the region ABC' would then yield information about the development of the motion before the camera was operated.

These ideas and their generalization to cases where the characteristic curves for hyperbolic equations are not straight and the slope at a point even depends on the solution there are important in such topics as the steady supersonic flow of gases.

The behaviour of the solution described by Fig. 2.8 is important also in relation to the numerical method of solving the wave equation by using the corresponding difference equation. We are not concerned with the circumstances under which the solution of the difference equation tends to that of the differential equation, a topic which belongs to the field of numerical analysis, but we indicate how the formal calculation of the solution leads to similar conclusions with regard to its determinacy as those already derived for the differential equation.

When a function of x is given numerically, its value is specified at a sequence or 'mesh' of x-values which for simplicity we shall take to be equally spaced along the x-axis. In the case of two independent variables x and t the mesh consists of a two-dimensional 'grid' in the (x, t) plane, drawn in Fig. 2.9. We define difference operators $\Delta_x u$ and $\Delta_t u$ by

$$\Delta_x u(x, t) = \frac{u(x + h, t) - u(x, t)}{h}, \quad \Delta_t u(x, t) = \frac{u(x, t + k) - u(x, t)}{k},$$

$$(2.28a, b)$$

where h and k are the respective mesh sizes in the x and t coordinate

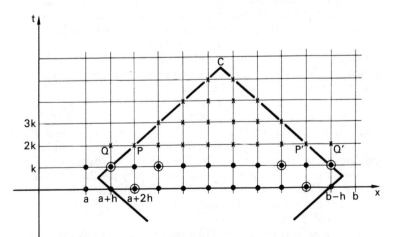

Figure 2.9

directions. Since $\lim_{h \to 0} \Delta_x u = \partial u / \partial x$ and $\lim_{k \to 0} \Delta_t u = \partial u / \partial t$, where the derivatives exist, the smaller the mesh size the more accurate it is to replace the partial derivatives $\partial u / \partial x$ and $\partial u / \partial t$ by (2.28). The aim is to replace the differential equation by a difference equation and for this purpose we define second difference operators by

$$\Delta_{xx} u(x, t) = \Delta_x [\Delta_x u(x - h, t)]$$
$$= \frac{u(x + h, t) - 2u(x, t) + u(x - h, t)}{h^2}, \tag{2.29}$$

$$\Delta_{tt} u(x, t) = \Delta_t [\Delta_t u(x, t - k)]$$
$$= \frac{u(x, t + k) - 2u(x, t) + u(x, t - k)}{k^2}. \tag{2.30}$$

It is left as an exercise for the reader to verify that the limits of these expressions as h and k tend to zero are $\partial^2 u / \partial x^2$ and $\partial^2 u / \partial t^2$ respectively.

In order to simplify the analysis we select the t-increment k to be h/c so that characteristic lines pass exactly through mesh points and then the result of replacing the derivatives in the wave equation by the approximate expressions (2.29) and (2.30) is a relation between values of u that may be written as

$$u(x, t + k) = u(x + h, t) - u(x, t - k) + u(x - h, t). \tag{2.31}$$

This formula is used to build up an array of values of u (a 'solution') in the upper half of the (x, t) plane at the mesh points. Consider the initial value problem discussed above in which $u(x, t)$ and $\partial u(x, t)/\partial t$ are pre-scribed on $t = 0$ for $a < x < b$; that is, at mesh points $x = a + h, a + 2h,$ ..., $b - h$. From (2.28b) the solution can be computed at $t = k$ so that $u(x, k)$ is known at the mesh points in $a < x < b$. The initial conditions are therefore equivalent to a double row of data on u at $t = 0$ and $t = k$. We may now calculate values of u at the later time $t = 2k$ by using equation (2.31). However, to find u at a given x requires a knowledge of u at $x - h$ and $x + h$ at the earlier time value. Thus the values of u at P and P' in Fig. 2.9 can be determined from the values at the encircled points but there is no way of calculating values of u at Q and Q'. The values of u along the whole row at $t = 2k$ can be filled in between P and P'. The process may be repeated for the rows $t = 3k, 4k, \ldots$ and it is evident that the range of x for which the solution can be found decreases as t increases and is bounded in the (x, t) plane by the characteristics through $x = a$ and $x = b$. The characteristic triangle ABC is thus the domain of dependence of the point C as we found earlier from con-sideration of the differential equation.

These ideas may be extended to more general problems although the numerical procedure may require modification. For an alternative treatment related to the approach to the solution via two first-order differential equations, discussed in Section 2.3, see Exercise 2.8 where the property of characteristics as lines along which combinations of the dependent variables have constant values is utilized to advance the solution. When applied to more general problems the characteristics themselves may not be known in advance; they can, however, be computed approximately, an increment at a time, hand-in-hand with the solution.

2.5 Wave Reflection and Transmission

The general solution of the wave equation representing a wave travelling to the right was derived in Section 2.3 to be $u = f(x - ct)$. Solutions where f is chosen as the sine or cosine of a multiple of $x - ct$ (e.g. $\cos[m(x - ct)]$, where m is a constant) are particularly useful. The reason for this is that because the equation is linear the Principle of Superposition may be applied as explained in Chapter 1. Hence, infinite sums (provided they are convergent and differentiable) of constant multiples of these 'sinusoidal' solutions, as they are called, also satisfy the wave equation and, by the method of Fourier series, sums of this type can be chosen to represent any desired function defined over a finite range. Thus, for many purposes, it is sufficient to consider a solution of the form,

$$u = a_1 \cos[2\pi k(x - ct)] + a_2 \sin[2\pi k(x - ct)]$$
$$= a \cos[2\pi k(x - ct) + \epsilon], \qquad (2.32)$$

where $a = \sqrt{(a_1^2 + a_2^2)}$, $\epsilon = \pi - \tan^{-1}(a_2/a_1)$, k is a constant and the factor 2π is included for convenience. A wave of this form, like that whose profile at $t = 0$ is drawn in Fig. 2.10a, is said to be *harmonic*. One important property of such waves is that they are periodic. Thus, at fixed t, the wave profile repeats itself every $1/k$ units of distance along the x-axis. This quantity $1/k$ is called the *wavelength* λ. It follows that the quantity k measures the number of wavelengths in unit distance and is known as the *wave number*. It has dimensions (length)$^{-1}$.

Analogous quantities can be introduced using the variation of u with time rather than distance. At a fixed x, say $x = 0$, (as in Fig. 2.10b) the value of u passes through the same cycle of values repeatedly as t changes. The minimum time it takes for a complete cycle to occur is known as the *period* of the wave, τ, where

$$\tau = \frac{1}{kc} = \frac{\lambda}{c}.$$

Since the wave travels with speed c, τ is the time it takes for a length λ of the wave to pass a fixed point. Corresponding to k we define the *frequency* n as the number of waves passing a fixed point in unit time. Thus $n = 1/\tau$ and has dimensions (time)$^{-1}$.

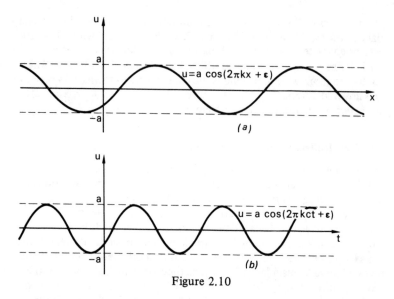

Figure 2.10

The *amplitude* of the wave is defined as the maximum value of u as x and t vary and is the quantity a in (2.32). It is a measure of the size of the disturbance. The constant ϵ in (2.32), the *phase constant* of the wave, is mainly of relative importance in comparing one wave with another; for example, at a given instant of time the maxima of u for two waves of the form (2.32) having the same wavelength, and phase constants ϵ_1 and ϵ_2 respectively, are separated by a distance $(\epsilon_1 - \epsilon_2)/2\pi k$. If $\epsilon_1 = \epsilon_2 + 2r\pi$, r an integer, the waves are said to be 'in phase' and if $\epsilon_1 = \epsilon_2 + (2r + 1)\pi$ they are 'exactly out of phase'. Intermediate cases correspond to a phase lead or phase lag of one wave relative to the other. The reader is recommended to sketch typical cases for himself.

The definitions introduced above apply also to the left-travelling harmonic wave, $u = a \cos[2\pi k(x + ct) + \epsilon]$.

A consequence of the linearity of the wave equation is that we can represent the solution (2.32) in the complex form,

$$u = A\, e^{2\pi i k(x - ct)}, \quad A = a\, e^{i\epsilon}, \tag{2.33}$$

since the real and imaginary parts separately are solutions. (As exponentials are generally simpler functions to handle than trigonometric ones the manipulations are more easily carried out.) It is understood that the physically significant quantities are to be obtained from this expression by taking the real part. However, the use of the complex form of the solution is only valid if linear operations are being applied. In other operations, such as squaring the solution, which is involved in determining the energy of the wave (see Section 2.8), the real part of (2.33) must be taken first.

Now right and left travelling harmonic waves, according to the results of Section 2.3, proceed unchanged so long as the medium remains uniform. However, in practice, there may arise changes in the medium as, for example, if strings of different density are joined together, or sound waves in an organ pipe reach a closed or open end. To investigate what happens in such cases we consider the simpler situation of a harmonic wave in $x < 0$ travelling to the right with the form,

$$u_i = A_i\, e^{2\pi i n_i(t - x/c_i)}, \tag{2.34}$$

where the suffix i denotes quantities associated with this incident wave and A_i is assumed real. We suppose that properties of the medium change discontinuously at $x = 0$ and some disturbance is communicated to the medium in $x > 0$ as a right-travelling wave,

$$u_t = A_t\, e^{2\pi i n_t(t - x/c_t)},$$

where the suffix t refers to the transmitted wave. Physical requirements at the discontinuity determine properties of the transmitted wave from those of (2.34) but it is easily verified that this simple model cannot satisfy all the conditions; the incident wave cannot, in general, travel across $x = 0$ unimpeded. Some of its energy is reflected as a left-travelling harmonic wave in $x < 0$ in the form,

$$u_r = A_r\, e^{2\pi i n_r(t + x/c_i)},$$

where the suffix r denotes the reflected wave. A_t and A_r may be complex and then their arguments are the phase constants of the transmitted and reflected waves relative to the incident wave.

Thus the problem of wave reflection and transmission consists of relating solutions of (2.1) for $x < 0$ and $x > 0$,

$$(u)_{x<0} = A_i\, e^{2\pi i n_i(t - x/c_i)} + A_r\, e^{2\pi i n_r(t + x/c_i)},$$

$$(u)_{x>0} = A_t\, e^{2\pi i n_t(t - x/c_t)}, \tag{2.35a, b}$$

by physical conditions which hold at $x = 0$ for all t.

In the absence of applied forces at $x = 0$ these conditions may be written as

$$\lim_{x \to 0-} L_1 u = \lim_{x \to 0+} L_2 u, \quad \text{for all } t, \tag{2.36}$$

where L_1 and L_2 represent linear and homogeneous operations, involving u itself or its derivatives. On substituting the solutions (2.35) into (2.36), algebraic equations for A_r and A_t are obtained whose coefficients depend exponentially on t. Since (2.36) holds for all t the frequencies n_r and n_t are the same as n_i and we shall henceforth use n. We illustrate the method by giving some examples.

2.5.1 Density Discontinuity on a String

For the first example we consider the case of two uniform strings of different densities per unit length, ρ_1 and ρ_2, joined together at $x = 0$. Hence, by (2.4), the wave velocity is different in the two strings. According to the arguments just given we represent the transverse displacement of the string z by

$$z = \begin{cases} A_i\, e^{2\pi i n(t - x/c_i)} + A_r\, e^{2\pi i n(t + x/c_i)}, & x < 0, \\ A_t\, e^{2\pi i n(t - x/c_t)}, & x > 0, \end{cases} \tag{2.37}$$

where $c_i = \sqrt{(T_1/\rho_1)}$, $c_t = \sqrt{(T_2/\rho_2)}$, T is the tension and the suffixes 1 and 2 here and in Sections 2.5.2 and 2.5.3 refer to values in the domains $x < 0$ and $x > 0$ respectively.

The conditions that are applied at $x = 0$ to link these solutions are, in the first place, that z is continuous because the string remains unbroken, or

$$\lim_{x \to 0-} z(x, t) = \lim_{x \to 0+} z(x, t), \quad \text{for all } t. \tag{2.38}$$

Secondly, because there is no finite mass situated at $x = 0$, in order to avoid infinite accelerations there the net force due to the tensions in the right- and left-hand strings must balance. These tensions act tangentially to the string and hence the slope of the string is continuous at $x = 0$,

$$\lim_{x \to 0-}\left(\frac{\partial z}{\partial x}\right) = \lim_{x \to 0+}\left(\frac{\partial z}{\partial x}\right), \quad \text{for all } t. \tag{2.39}$$

The continuity of the magnitude of the tensions implies that we can write $T_1 = T_2 = T$, say.

On substituting the expressions (2.37) into (2.38) and (2.39) we obtain

$$A_i + A_r = A_t, \quad A_i - A_r = \frac{c_i}{c_t}\, A_t, \tag{2.40}$$

from which the amplitudes of the reflected and transmitted waves are obtained as

$$A_r = \left(\frac{1 - c_i/c_t}{1 + c_i/c_t}\right)A_i, \quad A_t = \frac{2A_i}{1 + c_i/c_t}.$$

In terms of the string densities these expressions are

$$A_r = \left(\frac{\sqrt{\rho_1} - \sqrt{\rho_2}}{\sqrt{\rho_1} + \sqrt{\rho_2}}\right)A_i, \quad A_t = \frac{2\sqrt{\rho_1}}{\sqrt{\rho_1} + \sqrt{\rho_2}}\, A_i. \tag{2.41}$$

Thus A_r and A_t are real and since $A_t > 0$ the transmitted wave is in phase with the incident wave but the reflected wave is so only when the right-hand string is less dense. Otherwise, if $\rho_2 > \rho_1$, $A_r < 0$ and as z_r may be written in the form, $z_r = |A_r| e^{2\pi i n(t + x/c_i) + i\pi}$, they are exactly out of phase.

If $\rho_1 = \rho_2$ there is no discontinuity and the solution (2.41) yields the correct limiting behaviour of the incident wave being transmitted unimpeded across $x = 0$.

2.5.2 Discontinuity in Depth of a Canal

An analogous situation arises in the case of gravity waves on shallow water (Section 2.2.4) when the mean depth changes discontinuously from h_1 to h_2 at $x = 0$. From (2.17) the wave speed differs in $x < 0$ and $x > 0$ and we assume solutions like (2.37) for Z, the displacement of the water surface from its undisturbed level, and c_i and c_t are $\sqrt{(gh_1)}$ and $\sqrt{(gh_2)}$.

Here the pressure is continuous at $x = 0$ for all t and hence, by (2.14), Z satisfies (2.38). Also, the conservation of mass in a small volume of water containing the plane $x = 0$ (Fig. 2.11) requires that $v(h + Z)$ is continuous for all t at $x = 0$, where v is the velocity. On neglecting Z compared with h, as in deriving (2.17), and differentiating with respect to t we obtain the condition that $h\, \partial v/\partial t$ is continuous at $x = 0$, or, from (2.15),

$$\lim_{x \to 0-} \left(h_1 \frac{\partial Z}{\partial x} \right) = \lim_{x \to 0+} \left(h_2 \frac{\partial Z}{\partial x} \right), \quad \text{for all } t. \tag{2.42}$$

The equations for A_t and A_r, derived from (2.38) and (2.42), are identical with equations (2.40) except that the role of the wave speeds

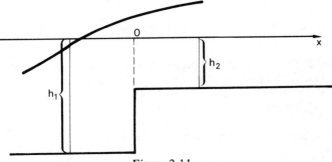

Figure 2.11

in the two regions is reversed. The solution is (2.41) with ρ_1 replaced by h_1 and ρ_2 by h_2. In contrast with the strings, the reflected and incident waves are in phase if they travel more rapidly than the transmitted wave.

2.5.3 Sound Waves in a Pipe

Here the incident wave is a sound wave moving in a straight pipe parallel to the x-axis and of constant cross-sectional area S. At $x = 0$ there is a moveable, thin rigid piston of mass m, assumed frictionless, and able to slide along the pipe (Fig. 2.12). The undisturbed gas is at rest and has uniform properties. The governing equation for sound propagation in the pipe is (2.12), provided the wavelength is much larger than the pipe perimeter, and again we assume the form (2.37) for, say, the particle velocity v with $c_i = c_t = c$.

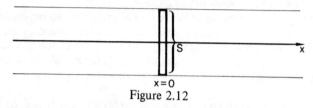

$$x = 0$$
Figure 2.12

As a result of the motion of the piston, induced by the incident wave, the gas motion in $x < 0$ is communicated to that in $x > 0$. However, certain conditions hold at the piston and these determine the properties of the transmitted and reflected waves.

An important simplification is that we can suppose the piston to be *fixed* at $x = 0$ in applying these conditions. The reason is that since the amplitude of the piston motion like that of the gas is assumed small any error incurred by replacing a function of x at the actual piston position by its value at $x = 0$ is, by a Taylor series expansion, of order at most equal to the amplitude and therefore negligible.

Since the waves are longitudinal, v is continuous across the piston for all t. Also, Newton's Law for the piston, moving under the influence of the pressure forces over its two flat faces, yields

$$Sp_0 \left[\lim_{x \to 0-} \bar{p}_1 - \lim_{x \to 0+} \bar{p}_2 \right] = m \lim_{x \to 0} \left(\frac{\partial v}{\partial t} \right), \quad \text{for all } t, \quad (2.43)$$

where $\bar{p}$ is the dimensionless pressure perturbation due to the wave motion. The relation between pressure and velocity is given by, $\partial \bar{p}/\partial t = \gamma\, \partial \bar{\rho}/\partial t = -\gamma\, \partial v/\partial x$, from (2.11), (2.10) and (2.8). Hence, from (2.37) for v_1 and v_2, we obtain

$$\bar{p}_1 = \frac{\gamma}{c} \left\{ A_i\, e^{2\pi i n(t-x/c)} - A_r\, e^{2\pi i n(t+x/c)} \right\},$$

$$\bar{p}_2 = \frac{\gamma}{c} A_t\, e^{2\pi i n(t-x/c)}.$$

In performing the integration with respect to t to obtain $\bar{p}$ we have set the arbitrary functions of x equal to zero. This is required by equation (2.10), which limits these arbitrary functions to constants, and condition (2.43) itself, which says the constant must be the same in each region.

Finally, a constant in the entire pressure field may be ignored.

On substituting the expressions for v and $\bar{p}$ into the conditions at $x = 0$, we obtain

$$A_i + A_r = A_t, \quad A_i - A_r = (1 + iq)A_t,$$

where $q = 2\pi n c m / S p_0 \gamma$. The solution is

$$\frac{A_r}{A_i} = \frac{q\, e^{-i[\pi/2 + \tan^{-1}(q/2)]}}{2\sqrt{(1 + q^2/4)}}, \quad \frac{A_t}{A_i} = \frac{e^{-i\tan^{-1}(q/2)}}{\sqrt{(1 + q^2/4)}}.$$

Thus the transmitted wave lags behind the incident wave by an amount $\tan^{-1}(q/2)$ and the reflected wave does so by an additional amount $\pi/2$.

It is often worth verifying that 'obvious' limiting cases may be recovered from general results. Thus the limiting values of A_r and A_t as the mass of the piston tends to zero, that is $q \to 0$, are $A_r = 0$, $A_t = A_i$. The incident wave is unimpeded as we should expect.

Also a rigid partition across the tube corresponds to a very heavy piston for which $q \to \infty$. Hence, $\tan^{-1}(q/2) \to \pi/2$ and we obtain the limiting values, $A_r = -A_i$, $A_t = 0$. There is total reflection of the incident wave with the reflected wave exactly out of phase with it. No wave is transmitted.

The energy of reflected and transmitted waves is considered in Section 2.8.

2.6 Boundary Conditions

So far in this chapter we have been mainly concerned with solutions of the wave equation in a medium which is supposed infinite in extent. In the next two sections we consider the influence on the solution, and consequent wave motion, of boundaries of the medium at fixed finite points on the x-axis. At such boundaries the solution u of the wave equation is required to satisfy prescribed conditions. In practice they are determined by physical requirements about the behaviour of the quantity represented by u (such as displacement of a string, pressure of a gas, etc.) at a boundary.

There is a connection with the topic of wave reflection at a discontinuity treated in Section 2.5 where one boundary was so remote as to have no influence on the solution and the incident wave was assumed to 'come from infinity' from the left. By contrast now both boundaries, which we suppose are situated at $x = 0$ and $x = l$, have significant effects on the solution. In each case the boundary conditions to be imposed take the form of conditions on u which must be satisfied at $x = 0$ and $x = l$ for all t. Referring to Section 1.4, the data at the x-boundaries is of the type (1.19); here, and in Section 2.4, the data at the t-boundary is of Cauchy type.

It is frequently possible to arrange for the value of u at a boundary to behave in any prescribed manner. For example, the displacement of an end of a stretched string can be made to have a desired variation with time by arranging for the point to which the string is attached to be suitably moved by external means. The motion of the string so produced is referred to as 'forced' since it continues as long as the external source is maintained and would not eventually die out as a result of the presence of friction. Energy is being fed continuously into the system through the boundary. We shall not discuss such cases. Here we consider *free oscillations* for which the source of disturbance is provided by the initial conditions only and thereafter no forces act to maintain the motion which eventually dies away due to the friction in the system. In mathematical terms this means that the boundary conditions, as well as the equation, are homogeneous in u (but not the initial conditions since otherwise D'Alembert's solution (2.21) would give $u \equiv 0$ as a solution satisfying all the conditions of the problem).

Examples of boundary conditions that are commonly met for some of the systems discussed in Section 2.2 are given below.

(a) In the familiar case of the flexible string with fixed ends the displacement z satisfies, $z(0, t) = z(l, t) = 0, t \geqslant 0$.

(b) For sound waves moving within and parallel to the axis of a straight pipe, as considered in Section 2.2.3, there are two principal cases to consider. First, if the pipe has a closed, rigid end at $x = 0$, the gas particles cannot move there so that the boundary condition is $v(0, t) = 0, t \geqslant 0$.

On the other hand if the pipe has an open end at $x = 0$ the condition is that the pressure there remains at its atmospheric value throughout the motion, or, $\bar{p}(0, t) = 0, t \geqslant 0$. In fact this result is not strictly true since the air outside the tube near $x = 0$ is moving and the condition is equivalent to neglecting the mass of this air. It can be improved by including a small end correction.

(c) For water waves in a tank the condition at a rigid end of the tank at $x = 0$ is that the water velocity vanishes there, $v(0, t) = 0, t \geqslant 0$.

(d) For the torsional waves in a rod discussed in Section 2.2.5, the condition at a clamped end at $x = 0$ is that the angular displacement is zero there, $\phi(0, t) = 0, t \geqslant 0$, whereas if the end is free the torque must vanish and the boundary condition is $\partial\phi/\partial x = 0, x = 0, t \geqslant 0$.

We shall incorporate some of these boundary conditions in examples and others are required for the exercises. First, however, we describe a method for solving the wave equation for the finite systems under discussion.

2.7 Solution of the Wave Equation in Finite Regions

In this section a method is described of obtaining solutions of the wave equation subject to given initial conditions, and boundary conditions at $x = 0$ and $x = l$. First we note that D'Alembert's solution (2.21) applies generally and can be used to obtain the solution for the case of a finite medium considered here (Exercise 2.20). However, we describe a different method of solution which has the advantage of not being restricted to the wave equation (it is applied in Chapters 3 and 4 where D'Alembert's solution cannot be used) and of giving rise to an instructive physical picture of how the motion is built up.

In mathematical terms the problem at hand is a solution of (2.1), subject to the initial conditions,

$$u = r(x), \quad \frac{\partial u}{\partial t} = s(x) \quad \text{at } t = 0, \quad 0 \leqslant x \leqslant l, \tag{2.44}$$

and the boundary conditions, for which we shall assume the simplest form,

$$u = 0 \quad \text{at } x = 0 \quad \text{and} \quad x = l, \quad \text{for } t \geqslant 0. \tag{2.45}$$

These equations describe, for example, the motion of a string with both ends held fast or the motion of air in an organ pipe which has both ends closed.

In order to simplify the problem we restrict the class of functions to which the solution belongs. To be specific we look for a solution in a form in which the independent variables are separated. Hence we write for the solution,

$$u(x, t) = X(x)T(t), \tag{2.46}$$

where X and T are unknown functions to be found. Although this may seem rather a severe restriction on the range of solutions that might be obtained we recall from Chapter 1 that the Principle of Superposition enables more general solutions to be obtained by adding together different solutions of the type (2.46). In fact, as we shall see, it imposes no practical limitation at all.

We may write the result of substituting the expression (2.46) into the differential equation (2.1) in the form,

$$\frac{1}{X} \frac{d^2 X}{dx^2} = \frac{1}{c^2 T} \frac{d^2 T}{dt^2}.$$

Now since X is independent of t, the left-hand side can depend only on x and similarly the right-hand side can depend only on t. As x and t are independent variables the only way in which a function of x can be made equal to a function of t for all x and t (in the appropriate ranges) is for both functions separately to be equal to the same constant. It turns out that the constant must be negative (Exercise 4.15) and we denote it by $-m^2$. Then the variables X and T satisfy the equations,

$$\frac{d^2 X}{dx^2} + m^2 X = 0, \quad \frac{d^2 T}{dt^2} + m^2 c^2 T = 0.$$

These are just equations of simple-harmonic motion having general solutions for X and T,

$$X = A \cos mx + B \sin mx, \quad T = C \cos mct + D \sin mct. \quad (2.47)$$

A, B, C and D are arbitrary constants but three only are independent because in the actual solution for u, (2.46), constant factors of X and T can be 'collapsed' into a single constant. However, m is also arbitrary and hence the four conditions (2.44) and (2.45) are sufficient to determine four independent constants in the solution. From the boundary conditions (2.45) on $u = XT$ we have, $X(0) = X(l) = 0$, since the only other possibility is that T vanishes identically which is inconsistent with (2.44). The first of these conditions yields $A = 0$ and the second then gives $B \sin ml = 0$. Either $B = 0$ or $\sin ml = 0$. The former possibility is rejected since it again gives $u(x, t) \equiv 0$. We thus obtain, $ml = n\pi$, or $m = n\pi/l$, where n is a positive integer (Exercise 2.15). In other words, instead of being entirely arbitrary, m must be an integral multiple of π/l in order that the boundary conditions be satisfied. This, of course, is a considerable curtailment of its range of values.

Before attempting to satisfy the initial conditions let us assess the solution as it stands so far. If we denote by u_n a solution corresponding to an integer n, we may write

$$u_n = \sin\left(\frac{n\pi x}{l}\right)\left[C_n \cos\left(\frac{n\pi ct}{l}\right) + D_n \sin\left(\frac{n\pi ct}{l}\right)\right], \quad (2.48)$$

where, without loss of generality, the constant B in X has been absorbed into C_n and D_n, whose values have still to be determined but in general depend on n. We can write this expression for u_n in a different form by introducing the amplitude A_n and a phase constant ϵ_n, defined by the equations, $C_n = A_n \cos \epsilon_n$, $D_n = A_n \sin \epsilon_n$, from which we obtain, $A_n = \sqrt{(C_n^2 + D_n^2)}$, $\tan \epsilon_n = D_n/C_n$, $0 \leqslant \epsilon_n < 2\pi$, with the proper quadrant determined by the sign of C_n and D_n. Equation (2.48) then becomes

$$u_n = A_n \sin\left(\frac{n\pi x}{l}\right)\cos\left(\frac{n\pi ct}{l} - \epsilon_n\right). \quad (2.49)$$

Each one of the infinite set of solutions represented by (2.48) or (2.49) satisfies the boundary conditions and is known, by analogy with systems of a finite number of degrees of freedom, as a *normal mode* of oscillation because it has the property that in such a motion all parts of the system oscillate at the same *normal frequency*, $nc/2l$. These normal mode solutions are important in analysing the general solution of the wave equation and play a major role in explaining the operation and properties of many musical instruments. From (2.49) we see that as time varies the wave profile does not move to the right or left but simply alters its amplitude. For this reason the solution is called a *standing wave* to distinguish it from the travelling waves previously considered.

The displacement of the string, for example, in a normal mode has the property that at a fixed x it oscillates simple-harmonically at the normal frequency. Certain points of the system, known as *nodes*, are never in motion. These occur where $\sin(n\pi x/l) = 0$ and, when $n = 1$, are at $x = 0$ and $x = l$, which are in fact nodes for all n to satisfy the boundary conditions. In general there are $n + 1$ nodes for u_n and typical shapes of a string vibrating in normal modes for which $n = 1, 2, 3, 4$, are drawn in Fig. 2.13. The wavelengths are $2l, l, 2l/3, l/2$, respectively.

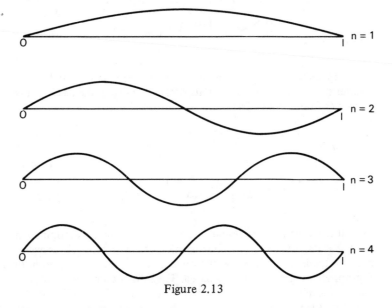

Figure 2.13

Now, as we shall see below, the actual solution of an initial value problem for the wave equation with the boundary conditions (2.45) is constructed from many of these different normal mode solutions. The special value $n = 1$ is termed the *fundamental mode* because the corresponding frequency, $c/2l$, determines the 'pitch' of a note emitted by a musical instrument. The values of $n > 1$ are referred to as 'higher harmonics' or 'overtones' and the relative magnitude of these modes identifies the quality or character of the note and, indeed, the instrument producing it, independently of either the note's pitch or loudness. The latter property depends essentially on the magnitude of the disturbance.

From (2.48), by superposition, we may obtain the more general solution,

$$u = \sum_{n=1}^{\infty} u_n = \sum_{n=1}^{\infty} \sin\left(\frac{n\pi x}{l}\right)\left[C_n \cos\left(\frac{n\pi ct}{l}\right) + D_n \sin\left(\frac{n\pi ct}{l}\right)\right],$$

$$(2.50)$$

and the question then is whether the doubly infinite set of arbitrary constants, C_n and D_n, can be chosen to satisfy

$$u(x, 0) = r(x), \quad \left[\frac{\partial u}{\partial t}\right]_{t=0} = s(x), \quad 0 \leqslant x \leqslant l.$$

If so, then,

$$\left. \begin{aligned} r(x) &= \sum_{n=1}^{\infty} C_n \sin\left(\frac{n\pi x}{l}\right), \\ s(x) &= \sum_{n=1}^{\infty} \left(\frac{n\pi c}{l}\right) D_n \sin\left(\frac{n\pi x}{l}\right), \quad 0 \leqslant x \leqslant l. \end{aligned} \right\} \qquad (2.51\text{a, b})$$

But from Section 1.6 this is precisely the problem of determining the Fourier sine series for the functions $r(x)$ and $s(x)$ and the coefficients C_n and D_n are, from (1.24),

$$\left. \begin{aligned} C_n &= \frac{2}{l} \int_0^l r(x) \sin\left(\frac{n\pi x}{l}\right) dx, \\ D_n &= \frac{2}{n\pi c} \int_0^l s(x) \sin\left(\frac{n\pi x}{l}\right) dx. \end{aligned} \right\} \qquad (2.52\text{a, b})$$

Once these integrals have been evaluated the solution is determined in the sense that its series is known. Furthermore, by Theorem 1.3, convergence can be established and, hence, the value of u for any x and t within their appropriate ranges may be calculated. The rate of convergence may be quite slow, unlike the analogous form of solution of the diffusion equation (see Chapter 3).

The solution (2.49) shows that individual normal modes can be excited if special initial conditions are applied. For example, if a stretched string is pulled into the shape $z(x, 0) = a \sin(N\pi x/l)$, for some fixed integer N and constant a, and released from rest the series solution reduces to a single term, $z(x, t) = a \sin(N\pi x/l) \cos(N\pi ct/l)$, using (2.52) with $s(x) = 0$ and $r(x) = a \sin(N\pi x/l)$.

We can check that the function represented by the series (2.48) with C_n and D_n determined by (2.52) does indeed satisfy the wave equation. We re-write (2.48) in the form,

$$u = \frac{1}{2} \sum_{n=1}^{\infty} \left\{ C_n \sin\left[\frac{n\pi}{l}(x - ct)\right] + D_n \cos\left[\frac{n\pi}{l}(x - ct)\right] + \right.$$

$$\left. + C_n \sin\left[\frac{n\pi}{l}(x + ct)\right] - D_n \cos\left[\frac{n\pi}{l}(x + ct)\right] \right\}. \qquad (2.53)$$

For values of x and t such that $0 \leqslant x - ct \leqslant l$ by (2.51a) the first series on the right-hand side has the sum, $r(x - ct)/2$, $0 \leqslant x - ct \leqslant l$. Hence using the results given in Section 1.6 the sums involving C_n in (2.53) are

$$\tfrac{1}{2}[r^*(x - ct) + r^*(x + ct)], \quad \text{all } x, t, \tag{2.54}$$

where $r^*(x)$ is the odd, periodic extension of $r(x)$, defined in $0 \leqslant x \leqslant l$. Similarly, after differentiating the series containing D_n with respect to t, assuming this step to be valid, we obtain

$$\frac{\partial}{\partial t} \left\{ \frac{1}{2} \sum_{n=1}^{\infty} D_n \left[\cos \left[\frac{n\pi}{l} (x - ct) \right] - \cos \left[\frac{n\pi}{l} (x + ct) \right] \right] \right\}$$

$$= \tfrac{1}{2}[s^*(x - ct) + s^*(x + ct)], \quad \text{all } x, t$$

where $s^*(x)$ is the odd, periodic extension of $s(x)$. The sum of the required series is obtained, on integration with respect to t, as,

$$\frac{1}{2c} \int_{x-ct}^{x+ct} s^*(\sigma) \, d\sigma, \quad \text{all } x, t. \tag{2.55}$$

The complete solution (2.50) can thus be expressed as the sum of (2.54) and (2.55) and is valid for all x and t. It is recognizable as D'Alembert's solution in the form (2.25). Moreover, this result shows that even when the disturbance is confined to a finite region of the x-axis we can still regard the solution as being built up of waves travelling to the right and left without change of shape provided that the definition of the functions describing the initial state of the medium (for example, the initial displacement and velocity of a string) is extended in the proper way.

We end this section by giving two examples.

Example (i). A plucked string fixed at $x = 0$ and $x = l$.

In this case, the string is initially pulled into some displaced position and released from rest. If $z(x, t)$ is the general value of the displacement, $\partial z/\partial t(x, 0) = s(x) = 0$, and hence the D_n in (2.50) all vanish. Suppose the string is displaced a distance a at its mid-point at $t = 0$. Then,

$$z(x, 0) = r(x) = \begin{cases} 2ax/l, & 0 \leqslant x \leqslant l/2, \\ 2a(l - x)/l, & l/2 \leqslant x \leqslant l. \end{cases}$$

The coefficients C_n of the Fourier sine series are, from (2.52a),

$$C_n = \frac{2}{l} \int_0^l r(x) \sin \left(\frac{n\pi x}{l} \right) dx = \frac{2}{l} \left\{ \int_0^{l/2} \frac{2ax}{l} \sin \left(\frac{n\pi x}{l} \right) dx + \right.$$

$$\left. + \int_{l/2}^l \frac{2a(l - x)}{l} \sin \left(\frac{n\pi x}{l} \right) dx \right\},$$

and by routine integration by parts we obtain, $C_n = 8a \sin(n\pi/2)/n^2\pi^2$. Only the C_n with odd suffices are non-zero so we introduce a new summation index, m, defined by $n = 2m + 1$. Then the solution (2.50) is

$$z = \sum_{m=0}^{\infty} (-1)^m \frac{8a}{\pi^2(2m+1)^2} \sin\left[\frac{(2m+1)\pi x}{l}\right] \cos\left[\frac{(2m+1)\pi ct}{l}\right],$$

containing only odd harmonics, as we might have expected as the string's motion is symmetric about the point $x = l/2$.

The conditions in this problem have something in common with the operation of stringed instruments such as a harp or guitar. The pitch of the note, that is the frequency of the fundamental mode, is for strings proportional to the square root of the tension and inversely proportional to the length and the square root of the mass per unit length. This is known as Mersenne's Law. In the case of bowed instruments, however, such as a violin the situation is different in that a force acts throughout the time a note is being emitted (during the 'bowing'). This means that the oscillations are not free and an analysis should include an extra term in the original wave equation to represent the force.

Example (ii).   An organ pipe with one end closed and one end open.

A model for the organ pipe is a tube of length l with boundary conditions $v = 0$ at $x = 0$ and $\bar{p} = 0$ at $x = l$ for all t (Section 2.6). We represent the solution in the form of pressure waves so that the variable u in (2.1) now denotes $\bar{p}$. Since $v = 0$ at $x = 0$ for all t it follows that $\partial v/\partial t = 0$ at $x = 0$ and therefore from (2.9), $\partial \bar{p}/\partial x = 0$ is the boundary condition on $\bar{p}$ at $x = 0$.

Hence the constants A and B in the solution (2.47) for $X(x)$ must here be chosen to satisfy $dX/dx = 0$ at $x = 0$, $X = 0$ at $x = l$. The first of these conditions yields $B = 0$ and the second requires that $A \cos ml = 0$, or $ml = (2n + 1)\pi/2$, $n = 0, 1, 2, \ldots$. The general solution in this case is therefore

$$\bar{p} = \sum_{n=0}^{\infty} \cos\left[\left(\frac{2n+1}{2}\right)\frac{\pi x}{l}\right]\left\{C_n \cos\left[\left(\frac{2n+1}{2}\right)\frac{\pi ct}{l}\right] + \right.$$

$$\left. + D_n \sin\left[\left(\frac{2n+1}{2}\right)\frac{\pi ct}{l}\right]\right\}, \tag{2.56}$$

where C_n and D_n are to be determined from initial conditions.

The normal mode solutions $\cos[(2n + 1)\pi x/2l]$ have the property that they all vanish at the open end and have maxima at the closed end. The fundamental mode $(n = 0)$ and the higher harmonics, $n = 1$ and $n = 2$, are drawn in Fig. 2.14. The fundamental frequency is here $c/4l$ so that the pitch of a note emitted by the pipe depends on the length of pipe and the speed of sound in the gas.

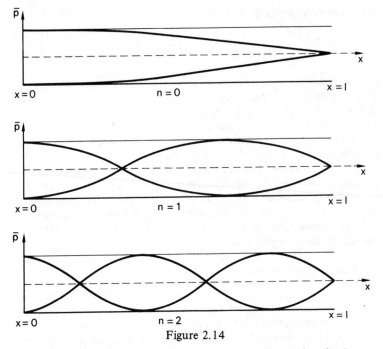

Figure 2.14

If the gas is initially at rest with a prescribed pressure distribution, then the initial conditions are $\bar{p} = P(x)$, $\partial \bar{p}/\partial t = -\gamma\, \partial v/\partial x = 0$, at $t = 0$, $0 \leqslant x \leqslant l$. The constants D_n all vanish from the second of these conditions and from the first the C_n must then satisfy

$$P(x) = \sum_{n=0}^{\infty} C_n \cos\left[\left(\frac{2n+1}{2}\right)\frac{\pi x}{l}\right], \quad 0 \leqslant x \leqslant l.$$

This is not a Fourier cosine series in the standard form but it can be considered as one over the double interval $0 \leqslant x \leqslant 2l$ provided the definition of $P(x)$ is extended in an anti-symmetric manner for $l \leqslant x \leqslant 2l$. The modifications are discussed in Chapter 1, Exercise 1.13, and the coefficients C_n are

$$C_n = \frac{2}{l} \int_0^l P(x) \cos\left[\frac{(2n+1)}{2}\frac{\pi x}{l}\right] dx.$$

For a given $P(x)$ this completes the solution.

As a simple illustration if the initial dimensionless pressure distribution is uniform and equal to P_0, the coefficents C_n are

$$C_n = \frac{(-1)^n 4P_0}{\pi(2n+1)},$$

and the pressure in the pipe at any subsequent time is, from (2.56),

$$\bar{p}(x,\,t) = \frac{4P_0}{\pi} \sum_{n=0}^{\infty} \frac{(-1)^n}{(2n+1)} \cos\left[\left(\frac{2n+1}{2}\right)\frac{\pi x}{l}\right] \cos\left[\left(\frac{2n+1}{2}\right)\frac{\pi c t}{l}\right].$$

In wind instruments the column of air in the tube is usually caused to vibrate by an exciting mechanism at one end arranged in various ways. Thus it is caused in an organ pipe by a jet of air across a lip, in a trombone or a trumpet by the instrumentalist's lips and in a clarinet or oboe by a reed. Again, however, the analysis of free oscillations of a tube of air is too simplified to describe the manner in which the note of the instrument is selected. This is because in each case the force which excites vibration is sustained for the duration of the note and the oscillations so produced are forced oscillations. Briefly, this implies that those normal frequencies of the system which are equal to the frequencies associated with the applied force are greatly enhanced. The instrument is said to resonate at these frequencies.

2.8 Energy Considerations

We now turn to an aspect of the wave equation which, whilst not concerned directly with its solution, is nevertheless important in leading to a physical understanding of the propagation phenomena. The variables satisfying (2.1), discussed in this chapter, represent systems in motion and hence we can define quantities to denote kinetic and potential (or electromagnetic) energies of these systems. For example, for the case of the string (Section 2.2.1) an element δx has mass $\rho_0\,\delta x$ and velocity $\partial z/\partial t$ and therefore its kinetic energy is, $\rho_0(\partial z/\partial t)^2\,\delta x/2$. For the finite length of string in $0 \leqslant x \leqslant l$ the kinetic energy is

$$\frac{\rho_0}{2} \int_0^l \left(\frac{\partial z}{\partial t}\right)^2 dx. \tag{2.57}$$

The potential energy that the string element possesses in its displaced position arises as a result of the work done against the tension in stretching it from its equilibrium length δx to its displaced length δs. Since the tension acts along the string this work is the product $T_0(\delta s - \delta x)$, or

$$T_0[(\delta x^2 + \delta z^2)^{1/2} - \delta x]$$

$$= T_0\,\delta x \left[1 + \frac{1}{2}\left(\frac{\partial z}{\partial x}\right)^2 + 0\left(\frac{\partial z}{\partial x}\right)^4 - 1\right] \approx \frac{T_0}{2}\left(\frac{\partial z}{\partial x}\right)^2 \delta x.$$

Again the integral,

$$\frac{T_0}{2} \int_0^l \left(\frac{\partial z}{\partial x}\right)^2 dx, \tag{2.58}$$

is needed for the string between $x = 0$ and $x = l$.

Analogous considerations apply in other cases. For water waves considered in Section 2.2.4, the mass of a column element of water of height $h + Z$ and length δx has kinetic energy per unit width, $\rho(h + Z)v^2 \, \delta x/2 = \rho h v^2 \, \delta x/2$, to leading order, where v is the water speed and ρ its density. Its potential energy per unit width measured relative to the mean height is that of a mass $\rho Z \, \delta x$ whose centre of gravity is at a height $Z/2$. Accordingly the expression for the potential energy of the element is $\rho g Z^2 \, \delta x/2$.

Finally, for sound waves the kinetic energy per unit area has the same form $\rho v^2 \, \delta x/2$. The potential energy of an elementary volume of gas whose mass per unit area is $\rho \, \delta x$ results from its being compressed. Its magnitude is conveniently calculated as the work done by the pressure forces in returning the volume to conditions in the undisturbed state. Only the perturbed part of the pressure $p_0 \bar{p}$ (see (2.8)) is effective and, for a volume per unit mass $1/\rho$, the work done per unit mass is

$$p_0 \int_{\bar{\rho}}^{\bar{\rho}=0} \bar{p} \, d\left(\frac{1}{\rho}\right) = -\frac{\gamma p_0}{\rho_0} \int_{\bar{\rho}}^{0} \bar{\rho} \, d\bar{\rho} = \frac{c^2 \bar{\rho}^2}{2},$$

on replacing $\bar{p}$ by $\gamma \bar{\rho}$ from (2.11) and $d(1/\rho)$ by $-d\bar{\rho}/\rho_0$ by (2.8). For a mass per unit area $\rho \delta x \approx \rho_0 \, \delta x$ the work done and therefore the potential energy per unit area is $\rho_0 c^2 \bar{\rho}^2 \, \delta x/2$.

We next consider the result of substituting D'Alembert's solution of the wave equation into these formulae. To be definite we shall treat the case of the string but similar results apply in other cases. With the displacement z given by (2.21) at any instant the length l of string between $x = 0$ and $x = l$ has

$$\text{kinetic energy} = \frac{\rho_0 c^2}{2} \int_0^l (f' - g')^2 \, dx,$$

$$\text{potential energy} = \frac{T_0}{2} \int_0^l (f' + g')^2 \, dx,$$

where dashes on f and g denote differentiation with respect to $x - ct$ and $x + ct$ respectively. If f or g vanishes, so that the wave motion reduces to one travelling to the left or right only, the kinetic and potential energies are equal since $T_0 = \rho_0 c^2$. Furthermore, by adding these expressions and writing $\rho_0 c^2 = T_0$, we obtain the result that the total energy is

$$T_0 \int_0^l (f'^2 + g'^2) \, dx = T_0 \int_0^l f'^2 \, dx + T_0 \int_0^l g'^2 \, dx.$$

Thus the total energies of the right- and left-travelling parts of a wave on a string are additive but not the kinetic and potential parts separately.

Waves can act also to transmit energy. The energy of water waves is dissipated due to their breaking on a beach which may be many miles from the storm that originally gave rise to the disturbance. The earth receives energy from the sun through the agency of electromagnetic waves. In general then we may expect a flow of energy to occur in wave motion and to see how this occurs we again fix attention on the elastic string and consider the simple case of a right-travelling sine wave of wave number k, $z = a[\sin 2\pi(kx - nt)]$, in Fig. 2.15. At the point A the string to the left exerts on the string to the right a force T_0 along the tangent.

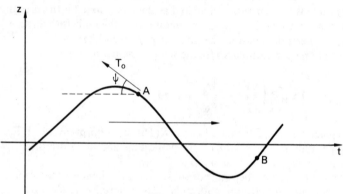

Figure 2.15

But with the wave moving to the right, in a small time interval δt, the particle of string at A, having speed v, moves a distance $v\,\delta t$ upwards (parallel to the z-axis). Hence the string to the left of A, in time δt, does an amount of work $T_0 v \sin\psi\,\delta t$ on the string to the right. Since $\sin\psi \approx \tan\psi = -\partial z/\partial x$ this work is

$$-T_0\frac{\partial z}{\partial x}\frac{\partial z}{\partial t}\delta t = T_0 a^2 4\pi^2 nk \cos^2[2\pi(kx - nt)]\delta t > 0.$$

The reader may verify that the same result holds at a point B where the slope of the string is positive. Thus the component of force always acts in phase with the particle velocity; any small element of string performs a positive amount of work on the element to its right and likewise receives a positive amount of work from the element to its left. In this way the wave acts as a transmitter of energy along the string. The rate at which this work is done is the power transmitted and is $4\pi^2 T_0 a^2 nk \times \cos^2[2\pi(kx - nt)]$, whose average value over a period from the fixed time t_0 to $t_0 + \tau$, is

$$\frac{1}{\tau}\int_{t_0}^{\tau + t_0} 4\pi^2 T_0 a^2 nk \cos^2[2\pi(kx - nt)]\,dt$$

$$= \frac{1}{\tau} \left\{ 4\pi^2 T_0 a^2 nk \int_{t_0}^{\tau + t_0} \frac{[1 + \cos 4\pi(kx - nt)]}{2} \right\} dt$$

$$= \rho_0 ca^2 n^2 2\pi^2. \tag{2.59}$$

Similar results hold for left-travelling waves.

As an illustration we return to the wave reflection problem at a density discontinuity in a string treated in Section 2.5. Since there is no loss or build-up of energy in the system the average rate at which energy is being transmitted by the incident wave must equal the sum of the corresponding average rates for the reflected and transmitted waves. Using formula (2.59) and referring to (2.41) this sum is

$$2\pi^2 \rho_1 c_1 A_r^2 n^2 + 2\pi^2 \rho_2 c_2 A_t^2 n^2$$

$$= 2\pi^2 A_i^2 n^2 \left\{ \frac{\rho_1 c_1 (\sqrt{\rho_1} - \sqrt{\rho_2})^2 + 4\rho_1 \rho_2 c_2}{(\sqrt{\rho_1} + \sqrt{\rho_2})^2} \right\}$$

$$= 2\pi^2 \rho_1 c_1 A_i^2 n^2,$$

as required. Similar energy relations may be verified in the other reflection examples that were considered.

The fractions of the average rate of flow of reflected and transmitted energies to that of the incident energy are defined as the *coefficients of reflection, R,* and *transmission, T*,* respectively. Because of the result just derived for the string the relation, $T^* = 1 - R$ holds and as the wave speed is the same for the incident and reflected waves, $R = |A_r|^2/|A_i|^2$, where the modulus signs are needed, in general, since A_r may be complex. Hence R also measures the fraction of the incident energy per unit length that is reflected but T^* has no such interpretation.

Next we notice that in a small time interval δt the energy carried by the waves which progress past A in Fig. 2.15 is, using the result that the kinetic and potential contributions are equal,

$$\rho_0 \int_0^{c \, \delta t} \left(\frac{\partial z}{\partial t} \right)^2 dx \approx 4\pi^2 \rho_0 n^2 a^2 c \, \delta t \cos^2 [2\pi(kx - nt)],$$

and its average value over a wavelength is

$$\frac{4\pi^2 \rho_0 n^2 a^2 c \, \delta t}{1/k} \int_0^{1/k} \cos^2 [2\pi(kx - nt)] \, dx = 2\pi^2 \rho_0 cn^2 a^2 \, \delta t.$$

This energy is being carried by the waves past A at a mean rate equal to that at which work is being done, (2.59). This means that for an isolated group of waves travelling along the string the speed at which energy is being transmitted is just sufficient to create new waves at the front of the group (and annihilate those at the back) that the group as a whole can travel with speed c. That this is not always the case is familiar to anyone who has thrown a stone into a pond and seen that the group of waves from the disturbance travels less rapidly than the individual waves within it.

Finally we consider the energy of a finite string fixed at $x = 0$ and l. We saw in Section 2.7 that the general solution is

$$z = \sum_{n=1}^{\infty} A_n \sin\left(\frac{n\pi x}{l}\right) \cos\left(\frac{n\pi ct}{l} - \epsilon_n\right), \tag{2.60}$$

using (2.49). The kinetic energy (K.E.) of the string is, from (2.57),

$$\text{K.E.} = \frac{\rho_0}{2} \int_0^l \left[\sum_{n=1}^{\infty} A_n \left(\frac{n\pi c}{l}\right) \sin\left(\frac{n\pi x}{l}\right) \sin\left(\frac{n\pi ct}{l} - \epsilon_n\right) \right]^2 dx. \tag{2.61}$$

Using the results that

$$\int_0^l \sin\left(\frac{n\pi x}{l}\right) \sin\left(\frac{m\pi x}{l}\right) dx = 0, \quad n \neq m,$$

$$\int_0^l \sin^2\left(\frac{n\pi x}{l}\right) dx = \frac{l}{2},$$

the product of the two infinite series in (2.61) is reduced to a single one and

$$\text{K.E.} = \frac{\pi^2 \rho_0 c^2}{4l} \sum_{n=1}^{\infty} n^2 A_n^2 \sin^2\left(\frac{n\pi ct}{l} - \epsilon_n\right). \tag{2.62}$$

Similarly, the potential energy (P.E.) of the string can be calculated from (2.58) using the solution (2.60). We obtain

$$\text{P.E.} = \frac{\pi^2 T_0}{4l} \sum_{n=1}^{\infty} n^2 A_n^2 \cos^2\left(\frac{n\pi ct}{l} - \epsilon_n\right). \tag{2.63}$$

On adding (2.62) and (2.63) we obtain the total energy of the string as

$$\frac{\rho_0 c^2 \pi^2}{4l} \sum_{n=1}^{\infty} n^2 A_n^2,$$

which is independent of time. Thus the total energy of the string is constant but the decomposition into potential and kinetic parts changes continuously throughout the motion. In practice friction causes the kinetic and potential energies to decrease and the motion to die away.

The expressions (2.62) and (2.63) have another point of interest. If we define an infinite set of coordinates ϕ_n by $\phi_n(t) = A_n \cos(n\pi ct/l - \epsilon_n)$, we can write

$$\text{K.E.} = \frac{\rho_0 l}{4} \sum \dot{\phi}_n^2, \quad \text{P.E.} = \frac{\pi^2 T_0}{4l} \sum_{n=1}^{\infty} n^2 \phi_n^2,$$

where $\dot{\phi}_n = d\phi_n/dt$. Thus, if we regard the string as a mechanical system of an infinite number of degrees of freedom with generalized coordinates ϕ_n, its K.E. and P.E. are sums of squares of $\dot{\phi}_n$ and ϕ_n. Since this is the property that normal coordinates have in vibrational problems for systems with finite numbers of degrees of freedom it can be helpful to regard the ϕ_n as normal coordinates. For example, in a normal mode of oscillation only one normal coordinate is non-zero as in the finite case.

EXERCISES

1. Derive the wave equation for the longitudinal vibration of a uniform bar placed parallel to the x-axis. Assume the material obeys Hooke's Law and consider the equations of motion of an element whose length in the unstressed equilibrium state is δx. Take as dependent variable the displacement u of a cross-section of the bar relative to its equilibrium position. The result is,

$$\frac{\partial^2 u}{\partial t^2} = c^2 \frac{\partial^2 u}{\partial x^2}, \quad c^2 = E/\rho,$$

where E is the Young's modulus and ρ the mass per unit length. What are the boundary conditions on u for (i) a clamped end, (ii) a free end, (iii) an end attached to a mass M which is free to move in the x-direction?

2. Suppose in the formula (2.25) the functions $r(x)$ and $s(x)$ are both defined for $a < x < b$. Show that the solution of the initial value problem for the wave equation is uniquely determined within the parallelogram bounded by the characteristics through $(a, 0)$ and $(b, 0)$ in the (x, t) plane.

3. Use the D'Alembert result (2.21) to show that if a function u is given along two intersecting characteristics $x - ct = a$, $x + ct = b$, from $(a, 0)$ and $(b, 0)$ respectively to the point of intersection, then $\partial^2 u/\partial t^2 = c^2 \partial^2 u/\partial x^2$ has a unique solution within the parallelogram defined in Exercise 2, but not elsewhere.

4. An example of the wave equation in which both independent variables are spatial arises in the theory of inviscid, supersonic flow past a thin two-dimensional body such as an aircraft wing of large span. In some conditions the velocity potential ϕ of the disturbance to the oncoming flow, assumed in the x-direction, relative to axes fixed in the wing, satisfies, $\partial^2\phi/\partial y^2 = \beta^2\partial^2\phi/\partial x^2$, where β is a constant. The boundary conditions at the wing surface, which to the order of accuracy of the theory can be taken to be at $y = 0$, are

$$\left(\frac{\partial \phi}{\partial y}\right)_{y=0+} = \frac{d\zeta_1}{dx}, \quad \left(\frac{\partial \phi}{\partial y}\right)_{y=0-} = \frac{d\zeta_2}{dx},$$

where $y = \zeta_1(x)$ and $y = \zeta_2(x)$ are the upper and lower surfaces of the wing, and for an unbounded medium ϕ satisfies conditions of zero disturbance at $x = -\infty$. Use D'Alembert's solution to find ϕ throughout the flow. (The perturbed pressure distribution over the wing is proportional to $\partial \phi / \partial x$ in this theory and from the solution important physical quantities such as the components of force on the wing in the x and y directions—the so-called drag and lift—can be calculated.)

5.   An infinite string has initial displacement given by $z(x, 0) = a\{1 - (x/b)^2\}$ for $|x| \leqslant b$ and $z(x, 0) = 0$ for $|x| > b$. Its initial velocity is zero. Use (2.25) to find the displacement for $t > 0$ and sketch the shape of the string at times $t = b/2c, b/c, 2b/c$.

6.   An infinite string has zero initial displacement but an initial velocity distribution given by $\partial z / \partial t = U$ (constant) for $|x| \leqslant b$ and $\partial z / \partial t = 0$ for $|x| > b$. Find the displacement for $t > 0$ and sketch the shape at times $t = b/2c, b/c, 2b/c$.

7.   The initial perturbed pressure distribution in an infinite vibrating column of air is $P \sin(x/b)$. For what choice of the initial velocity distribution does the resulting motion for $t > 0$ represent a travelling sound wave moving in the negative x direction?

8.   Regard the system $\partial u / \partial t = c \, \partial v / \partial x, \, \partial v / \partial t = c \, \partial u / \partial x$, where v is an additional new variable, as equivalent to the wave equation for u. By using the expressions for Δ_x and Δ_t obtain the difference equations for u and v and use the results that $u + v$ is constant on $x + ct = $ const. and $u - v$ is constant on $x - ct = $ const. to show that a knowledge of u and $\partial u / \partial t$ on the segment $a < x < b$ of the axis $t = 0$ is sufficient to determine u and v within the region bounded by the characteristics through a and b.

9.   The surface of water, whose undisturbed level lies along the x-axis at a height 0.1 m above the bottom, is in the form of a sine wave of amplitude 0.05 m and wavelength 0.5 m. At time $t = 0$ the surface of the water at $x = 0$ is moving downwards. Find the equation, period, frequency and wave number of the wave. What is the displacement and velocity of a point on the surface 0.75 m to the right of $x = 0$ at $t = 3/8$ secs? At this time sketch the shape of the water surface for a length 1 m to the right of $x = 0$. (Take g to equal 10 ms^{-2}.)

10.   A harmonic right-travelling wave of the form $z = A_i \, e^{2\pi i n(t-x/c)}$, ($A_i$ real), on an infinite uniform string is incident on a small mass m attached to the string at $x = 0$. Determine the conditions to be satisfied by the displacement at $x = 0$ and find the amplitude and phase of the

reflected and transmitted waves. Discuss the limiting cases as $m \to 0$ and $m \to \infty$. (Neglect gravity.)

11. Repeat the example of Section 2.5.3 for the case where the gases on the two sides of the piston have different undisturbed densities.

12. Find the reflection and transmission coefficients for a harmonic sound wave passing along a tube which has a discontinuity of cross-sectional area from S_1 to S_2 at $x = 0$.

13. Repeat Exercise 10 with the small mass m replaced by a light spring of stiffness k at $x = 0$. In the undisplaced position of the string the spring has its natural length. Discuss the limits $k \to 0$ and $k \to \infty$.

14. If a solution of the wave equation for the e.m.f. V in Section 2.2.3 is $V = f(x - ct) + g(x + ct)$, $c = (LC)^{-1/2}$, show that the current is $I = (C/L)^{1/2}\{f(x - ct) - g(x + ct)\}$.

A lossless transmission line ends at $x = l$ at a resistance R_l. If the incident e.m.f. is $V = f(x - ct)$ show that at $x = l$ the ratio of the reflected to the incident wave is $[R_l - (L/C)^{1/2}]/[R_l + (L/C)^{1/2}]$. Discuss the reflected voltage and current waves for the special cases of a short-circuit $(R_l \to 0)$ and an open circuit $(R_l \to \infty)$. [$(L/C)^{1/2}$ is called the characteristic impedance of the transmission line. When $R_l = (L/C)^{1/2}$ there is no reflected wave and the impedances are said to be '*matched*'.]

15. Verify in Section 2.7 that,

(i) a choice of separation constant $+m^2$ instead of $-m^2$ leads to a solution which cannot be made to satisfy all the conditions,

(ii) in the determination of $m = n\pi/l$ to satisfy the boundary condition $X(l) = 0$ there is no loss of generality in taking only positive integral values for n.

16. Initially the displacement of a string stretched between the fixed points $x = 0$ and $x = l$ is $z = a \sin(\pi x/l)$ and its velocity is $(bc/l) \sin(2\pi x/l)$. Find the displacement for all later times and show that there are certain values of t for which z vanishes for all x but there is no time at which the whole string is at rest.

17. A piano wire can be modelled by a uniform elastic string fixed at $x = 0$ and $x = l$. If the note is produced by the action of a hammer which imparts a uniform velocity U at $t = 0$ to the segment of wire $3l/8 \leqslant x \leqslant 5l/8$ whilst in its equilibrium position, calculate the displacement of the wire at any subsequent time.

18. Consider the flow in a pipe, as in Example (ii) of Section 2.7, for the following cases.

(i) The air in a closed pipe is raised to a pressure $1 + K$ times atmospheric and at $t = 0$ both ends are opened to the atmosphere.

(ii) Air is caused to flow through an open pipe with constant velocity V and at $t = 0$ both ends are closed.

In each case find solutions of the wave equation satisfying suitable initial and boundary conditions.

19.   The end $x = l$ of a rod of length l, rigidly clamped at $x = 0$, is twisted through an angle α, so that the cross-section at the station x is rotated through an angle $\alpha x/l$ from its position under no load, and then released at $t = 0$. Find the angle of twist at station x.

20.   Consider the application of D'Alembert's solution (2.21) to Example (i) in Section 2.7. To do this regard the finite string in $0 \leqslant x \leqslant l$ as a portion of an infinite string. Show that the solution on the infinite string which keeps $x = 0$ and $x = l$ fixed must be of the form $z = f(x - ct) - f(-x - ct)$, where the function $f(x)$ is odd and periodic with period $2l$, and that the initial conditions for the infinite string are the odd, periodic extensions of the initial conditions for the finite string in $0 \leqslant x \leqslant l$.
    Derive a simple graphical method of calculating the shape of the plucked string at any time by splitting the wave on the infinite string into its components moving to left and right. Interpret the result in terms of rules for reflection of a wave at a fixed end of a finite string.

21.   Consider two strings of different densities joined at $x = 0$ as in Section 2.5.1. Find the ratio of the densities of the strings if
   (i)   the amplitudes of the reflected and transmitted waves are equal,
   (ii)   the average power reflected is equal to that transmitted.
    Comment on the fact that there is only one possible answer to part (i) but two possible answers to part (ii).

22.   An infinite string is pulled into the shape $z = a \sin 2\pi k x$ and released from rest at $t = 0$. Calculate the shape at later times and find the average values of the kinetic and potential energies per unit length for a wavelength of the string. Discuss the fact that the sum of these energies is independent of time and relate it to the sum at $t = 0$.

23.   For Exercise 17 find an expression for the total energy of the piano wire and show that during the motion of the wire a fraction $8(2 - \sqrt{2})/\pi^2$ of the energy arises from the fundamental mode of vibration.

24.   Repeat Exercise 23 for the sound waves in the tube in Exercise 18 and torsion waves on the bar in Exercise 19.

25.   The bed of a channel is at $x = -h$ and a barrier is inserted in the channel so that the distribution of height of water is given by $z = 0$, $0 < x < l/2; z = a, l/2 < x < l$. The channel has closed ends at $x = 0$ and $x = l$. If the barrier is withdrawn at $t = 0$ find expressions for the kinetic and potential energies at any time and verify that their sum is equal to the total initial energy.

CHAPTER 3

The Diffusion Equation

3.1 Introduction

The experience of poking a fire and noting how the end of the poker away from the fire warms up is a familiar one. Heat from the fire is conducted along the poker and is felt by the hand. For a similar reason an effective method of cooling a hot cup of tea is to place in it a cold, metal tea-spoon for a short time. These are examples of a physical phenomenon which can be described in mathematical terms approximately by the one-dimensional diffusion equation,

$$\frac{\partial u}{\partial t} = \alpha^2 \frac{\partial^2 u}{\partial x^2}. \tag{3.1}$$

This equation is sometimes called the equation of heat conduction but we shall keep the term diffusion equation as being more descriptive of the general phenomena for which it serves as a mathematical model.

In most situations the independent variables x and t denote a space dimension and time as in the case of the wave equation, and the meaning to be attached to the dependent variable u depends on the particular physical problem being studied. The dimensions of the constant α^2, the *diffusivity* of whatever physical property u represents, are (length)^2/time and, unlike the wave equation, there is no velocity characteristic of (3.1). The equation itself possesses the property of smoothing out any differences there might be in u (for example, the temperature at the 'cold' end of the poker is raised); α^2 supplies a time scale for this smoothing process (see Section 3.5).

Equation (3.1) is the simplest example of a parabolic differential equation (Section 1.3) and like (2.1) it is linear. However, it contains only a first-order time derivative and this implies that we are not able to prescribe initial conditions on u and its first time derivative independently (Section 3.3). Consequently, there is a major difference in the behaviour of solutions of (3.1) from that of solutions of (2.1). To illustrate this we can compare the poker heated at one end with a finite string disturbed at one end, corresponding to the hot end of the poker, and fixed at the other. In the latter case, as we saw in Chapter 2, a wave propagates along the string from the disturbance, is reflected at the fixed end, and subsequently passes back and forth along the string. On the other hand, solutions of (3.1) do not represent waves and in the case of the poker the excess temperature at the hot end causes the heat to diffuse gradually through the solid with no oscillation of temperature at any point.

In Section 3.2, situations where the diffusion equation arises are described and typical initial and boundary conditions that can be imposed to obtain the solution are discussed in Section 3.3. The final two sections are concerned with properties of the equation and the application of separation of variables and other methods to obtain solutions which satisfy particular initial and boundary conditions.

3.2 Applications of the Diffusion Equation

3.2.1 Heat Conduction in a Solid

We shall derive the equation governing the heat flow in a solid in the full three-dimensional case as there is little complication in doing so and in any case we shall have cause to refer to it again in Chapter 4. The one-dimensional diffusion equation then follows as a special case.

The temperature field $\theta(x, y, z, t)$ is assumed to be continuous throughout the solid. It is known from experiment that q, the rate at which heat flows per unit area across a plane, is proportional to the normal temperature gradient across the plane and is directed towards the lower temperature; that is

$$q = -k \frac{d\theta}{dn},\tag{3.2}$$

where k is a positive quantity, the thermal conductivity of the medium, which we assume to be constant, and d/dn signifies differentiation along the normal.

Consider a volume element of solid, $\delta\tau$, drawn in Fig. 3.1, with rectangular faces whose edges are of length δx, δy and δz parallel to the coordinate axes. If the point P in Fig. 3.1 has coordinates (x_0, y_0, z_0), the faces lying in the planes $x = x_0$ and $x = x_0 + \delta x$ are denoted by A and B and similarly C, D and E, F denote the faces of $\delta\tau$ parallel to the planes $y = 0$ and $z = 0$ respectively. Thus, from (3.2), the rate at which heat flows into $\delta\tau$ through A can be written as $-k \overline{(\partial\theta/\partial x)}_A \, \delta y \, \delta z$, where $(\quad)_A$ denotes the average over A at time t. In time δt an amount of heat $-k \overline{(\partial\theta/\partial x)}_{A, \delta t} \, \delta y \, \delta z \, \delta t$ enters $\delta\tau$ through A, where the extra suffix δt means that the further average over the time δt is taken. Similarly an amount of heat $-k \overline{(\partial\theta/\partial x)}_{B, \delta t} \, \delta x \, \delta y \, \delta t$ leaves $\delta\tau$ in time δt through face B. Thus the net amount of heat entering $\delta\tau$ through faces A and B in time δt is,

$$\left\{ \left(\overline{\frac{\partial\theta}{\partial x}}\right)_{B, \delta t} - \left(\overline{\frac{\partial\theta}{\partial x}}\right)_{A, \delta t} \right\} k \, \delta y \, \delta z \, \delta t,$$

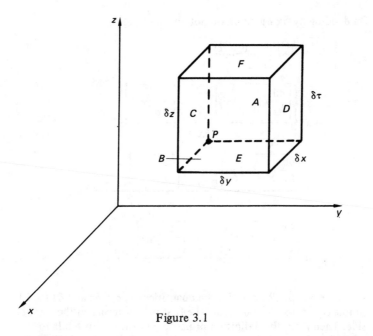

Figure 3.1

and similar contributions can be written down for the other two pairs of faces C, D and E, F.

Now the amount of heat contained in $\delta\tau$ at time t is $\rho c\,[\overline{\theta(t)}]_{\delta\tau}\,\delta x\,\delta y\,\delta z$ where $[\quad]_{\delta\tau}$ is the mean value in $\delta\tau$, and ρ and c are the density and heat capacity per unit mass of the solid. The heat gained by $\delta\tau$ in time δt can thus be expressed as

$$\rho c\{[\overline{\theta(t + \delta t)}]_{\delta\tau} - [\overline{\theta(t)}]_{\delta\tau}\}\,\delta x\,\delta y\,\delta z,$$

and provided there is no heat created or lost inside $\delta\tau$, as might be caused for example by a chemical reaction occurring in the solid, this must be equated to the total amount entering $\delta\tau$ in δt through its boundaries, namely,

$$\left\{\left(\overline{\frac{\partial\theta}{\partial x}}\right)_{B,\delta t} - \left(\overline{\frac{\partial\theta}{\partial x}}\right)_{A,\delta t}\right\} k\,\delta y\,\delta z\,\delta t +$$

$$+ \left\{\left(\overline{\frac{\partial\theta}{\partial y}}\right)_{D,\delta t} - \left(\overline{\frac{\partial\theta}{\partial y}}\right)_{C,\delta t}\right\} k\,\delta x\,\delta z\,\delta t +$$

$$+ \left\{\left(\overline{\frac{\partial\theta}{\partial z}}\right)_{F,\delta t} - \left(\overline{\frac{\partial\theta}{\partial z}}\right)_{E,\delta t}\right\} k\,\delta x\,\delta y\,\delta t.$$

On dividing by $\delta x \, \delta y \, \delta z \, \delta t$ we obtain

$$\left\{ \frac{[\theta(t + \delta t)]_{\delta\tau} - [\theta(t)]_{\delta\tau}}{\delta t} \right\}$$

$$= \alpha^2 \left\{ \frac{\left[\frac{\partial\theta}{\partial x}(x_0 + \delta x) \right]_{B, \delta t} - \left[\frac{\partial\theta}{\partial x}(x_0) \right]_{A, \delta t}}{\delta x} + \right.$$

$$+ \frac{\left[\frac{\partial\theta}{\partial y}(y_0 + \delta y) \right]_{D, \delta t} - \left[\frac{\partial\theta}{\partial y}(y_0) \right]_{C, \delta t}}{\delta y} +$$

$$\left. + \frac{\left[\frac{\partial\theta}{\partial z}(z_0 + \delta z) \right]_{F, \delta t} - \left[\frac{\partial\theta}{\partial z}(z_0) \right]_{E, \delta t}}{\delta z} \right\}, \tag{3.3}$$

where $\alpha^2 = k/\rho c$. We now let the quantities δx, δy, δz and δt tend to zero in this equation. Consider, say, the first pair of terms on the right-hand side. Then from the definition of partial derivative this tends to $\alpha^2 \, \partial^2\theta/\partial x^2$ provided that average values can be used in the definition. In fact the latter step is justified whenever the partial derivatives themselves are continuous functions and this we shall assume throughout our discussion. Thus from (3.3) we obtain the equation for heat flow in the solid in the form

$$\frac{\partial\theta}{\partial t} = \alpha^2 \left(\frac{\partial^2\theta}{\partial x^2} + \frac{\partial^2\theta}{\partial y^2} + \frac{\partial^2\theta}{\partial z^2} \right). \tag{3.4}$$

For this case α^2 is the thermal diffusivity.

For the special case of one-dimensional heat flow with which this chapter is concerned we must ensure that θ is independent of y and z. Such a situation is approximately realized for heat flow across a slab of infinite extent in the y and z directions but of finite thickness in the x-direction satisfying boundary conditions (Section 3.3) independent of y and z. Another example arises in the flow of heat along a uniform thin bar lying along the x-axis when the long faces are heat insulated. In both these cases we may apply the one-dimensional version of (3.4), $\partial\theta/\partial t = \alpha^2 \, \partial^2\theta/\partial x^2$.

3.2.2 Flow of Electricity in a Cable
As a second example we refer to the flow of electricity in a cable discussed in Section 2.2.3. There it was assumed that the resistance per unit length of cable, R, was negligible along with the conductance to earth

through the insulation. Here, however, we suppose that the effects of the cable's inductance are insignificant but we do take account of its resistance. This state of affairs is more appropriate in the operation of a submarine cable at audio-frequencies. Thus over a small element δx of cable the current change is still given by equation (2.13) but the potential change is now $\delta V = -IR \, \delta x$. On dividing by δx and letting δx tend to zero we obtain

$$\frac{\partial I}{\partial x} = -C \frac{\partial V}{\partial t}, \quad \frac{\partial V}{\partial x} = -IR. \tag{3.5}$$

Elimination of I and V in turn from these equations then gives the diffusion equation,

$$\frac{\partial u}{\partial t} = \alpha^2 \frac{\partial^2 u}{\partial x^2}, \quad \alpha^2 = (RC)^{-1},$$

for V and I.

3.2.3 Fluid Motion Near an Infinite Plane Wall

In certain circumstances the motion of a large volume of viscous fluid bounded by a plane wall is governed by the one-dimensional diffusion equation. Because of the action of viscosity (see Section 3.3) a movement of the wall in its own plane induces a motion of the fluid. We choose a fixed system of coordinates (x, y, z) such that the wall lies in the plane $z = 0$ and the fluid occupies the region $z > 0$. Then no physical property of the fluid, such as its velocity and pressure, can depend on the coordinates x and y. In other words, as is physically clear; at a given time all points in the fluid equidistant from the plane have the same motion.

The velocity components v_1, v_2 and v_3 in the x, y and z directions in an incompressible fluid satisfy the equation,

$$\frac{\partial v_1}{\partial x} + \frac{\partial v_2}{\partial y} + \frac{\partial v_3}{\partial z} = 0, \tag{3.6}$$

which, from the physical standpoint, expresses conservation of mass in the fluid. Under the present conditions the first and second terms vanish identically and hence $v_3 = v_3(t)$. Because the wall is solid the normal velocity component vanishes there and so we deduce that $v_3 \equiv 0$.

The momentum equations for the fluid must also be satisfied and they may be written as

$$\frac{\partial v_i}{\partial t} + \sum_{j=1}^{3} v_j \frac{\partial v_i}{\partial x_j} = -\frac{1}{\rho} \frac{\partial p}{\partial x_i} + v \sum_{j=1}^{3} \frac{\partial^2 v_i}{\partial x_j^2}, \quad i = 1, 2, 3, \tag{3.7}$$

where v is the kinematic viscosity, $(x_1, x_2, x_3) \equiv (x, y, z)$ and we have neglected the effect of any external forces such as gravity. On setting

$v_3 = 0$ and $\partial/\partial x = \partial/\partial y = 0$, these equations show that p is independent of position and v_1 and v_2 satisfy

$$\frac{\partial v_1}{\partial t} = \nu \frac{\partial^2 v_1}{\partial z^2}, \quad \frac{\partial v_2}{\partial t} = \nu \frac{\partial^2 v_2}{\partial z^2}.$$

In fact if the wall moves in a straight line, as we assume, we can choose the x-axis to point in that direction and then the zero boundary or initial conditions and uniqueness of the solution ensure that $v_2 \equiv 0$. Thus the velocity component of the fluid in the direction of the wall's motion satisfies the one-dimensional diffusion equation,

$$\frac{\partial u}{\partial t} = \nu \frac{\partial^2 u}{\partial z^2}, \quad \alpha^2 = \nu.$$

3.3 Initial and Boundary Conditions

It is convenient in the case of the diffusion equation to treat the initial and boundary conditions together. We have already said in Section 3.1 that one initial condition only is needed because the equation (3.1) contains only a first-order derivative with respect to time. This usually takes the form of information given on the dependent variable at $t = 0$ for the whole range of the space variable x. It may be helpful to introduce the idea of a steady state. This is a state of the physical system in which no changes with respect to time take place and indeed the field $u(x)$ in (3.1) then satisfies $d^2u/dx^2 = 0$. Such a state often acts as an initial condition for the unsteady problem that results when it is disturbed in some way.

For example, if an end of the poker of Section 3.1 is plunged into the fire the initial condition for the heating problem might be the steady state of constant temperature. When a long time has elapsed a new steady state is reached in which the temperature of the poker at any point is essentially independent of time (the actual value depending on the boundary conditions adopted); this state would then serve as the initial condition for the cooling problem on removal of the poker from the fire. Analogous situations arise for the other cases discussed in Section 3.2. For the viscous fluid the initial steady state is one of rest.

Since we have made no restriction on the size of the x-domain for which (3.1) holds, the boundary conditions may be applied at finite or infinite values of x. As usual they are determined by modelling physical requirements in a mathematical form.

For instance, if $\theta(x, t)$ is the temperature then one possibility is that θ has a known value at x corresponding to a boundary where the temperature is prescribed (by external means). Similarly, if a boundary is heat-insulated, so that no heat flows across it, then, according to (3.2), $\partial\theta/\partial x$ always vanishes there. The general case is described by Newton's Law of Cooling which says that the amount of heat radiated by a body at temperature θ into a medium of temperature Θ is proportional to $\theta - \Theta$. Thus, with h constant, the boundary condition takes the form,

$$\frac{\partial \theta}{\partial x} = -h(\theta - \Theta), \tag{3.8}$$

at the boundary for all t where Θ can depend on t. As $h \to \infty$ and $h = 0$ we recover the special cases of a boundary held at temperature Θ and an insulated boundary respectively. It is important to remember that such a condition must be satisfied at all boundaries even when infinitely far away if a unique solution is to be obtained (Section 1.5).

In the flow of electricity in a cable the values of the potential at the ends of the cable can be varied in a prescribed manner. A special case of this arises when one end is suddenly earthed so that its potential is subsequently zero.

Finally in the flow of a viscous fluid near a plane wall $z = 0$, discussed in Section 3.2.3, the velocity of the fluid at the wall is equal to that of the wall itself because of the physical requirement of no slipping between the fluid and a solid boundary. It follows that a wide variety of boundary conditions for the diffusion equation is applicable by suitably arranging the wall's velocity. In practice, there are two cases of particular importance, when the wall acquires a constant velocity impulsively from rest and when the wall oscillates sinusoidally at constant frequency. The fluid may extend to infinity in the z-direction in which case the velocity is assumed to vanish there or it may be bounded by a second wall parallel to $z = 0$ when the appropriate boundary condition must be applied there too.

3.4 Properties of The Diffusion Equation

The properties of the diffusion equation differ from those already discussed for the wave equation and from those to be treated in Chapter 4 for Laplace's equation. In a sense, as a parabolic equation, it occupies a position somewhere between the equations discussed in Chapters 2 and 4; this will emerge in what follows.

Although the approach via the method of characteristics, used successfully for the wave equation, is of limited value in the case of parabolic equations it does illustrate the mid-way position they have. We find it convenient to consider, instead of (3.1), the equation,

$$\frac{\partial^2 u}{\partial x^2} - \frac{1}{c^2} \frac{\partial^2 u}{\partial t^2} = \frac{1}{\alpha^2} \frac{\partial u}{\partial t}. \tag{3.9}$$

Thus the operator on u on the left-hand side of this equation is the one-dimensional wave operator and formally as the constant c, corresponding to the wave speed, tends to infinity equation (3.9) tends to the diffusion equation (3.1). The presence of the lower-order derivative on the right-hand side does not affect the calculation of the characteristics of the whole equation, as may be verified most easily by the method described at the end of Section 1.3. The characteristics of equation (3.9) are thus those of the wave equation drawn in Fig. 2.5. They are straight lines inclined at angles $\cot^{-1} c$ and $\pi\text{-}\cot^{-1} c$ with the x-axis. As c increases the characteristics 'open out' more and more so that in the limit as $c \to \infty$ they become parallel and anti-parallel to the x-axis respectively. This argument thus indicates that for the diffusion equation, obtained from (3.9) in this limit, only one family of characteristics, the lines $t = \text{const.}$, can be defined. Indeed, the property of possessing only a single family of characteristics is satisfied by parabolic partial differential equations in general.

A further property of (3.1) also emerges from the foregoing discussion. In contrast to the wave equation information is communicated instantaneously to all parts of the medium since the horizontal characteristic through a point P, say, in Fig. 3.2 carries the effect of any change in conditions at P (such as a sudden input of heat) to all relevant values of x at the same t-value.

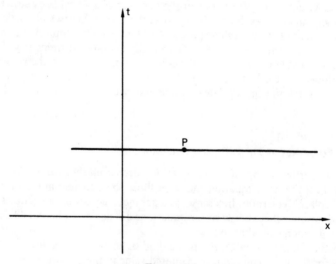

Figure 3.2

Further insight into the behaviour of the diffusion equation can be deduced from an examination of its associated difference equation. The required differences have already been defined by equations (2.28b) and (2.29). Here we choose for simplicity the x and t increments, h and k, to satisfy the relation $h^2 = \alpha^2 k$. Substitution of the approximations into equation (3.1) then yields the difference equation,

$$u(x, t + k) = u(x + h, t) - u(x, t) + u(x - h, t). \qquad (3.10)$$

First we note from this equation that only two time values are involved, namely t and $t + k$. This distinguishes it from the difference form (2.31) for the wave equation where information at the two previous time values was needed in order to proceed. It corresponds to the fact that for equation (3.1) both u and $\partial u/\partial t$ cannot be prescribed independently at $t = 0$. However, as in Section 2.4, the solution may still be advanced in the direction of the t-axis (that is to say the solution u can be obtained at neighbouring, increasing t-values successively), and we shall see that this property, which holds for (2.1) and (3.1), does not hold for the equation treated in Chapter 4. For a given row of data at some constant time T (such as that in Fig. 3.3 at $t = 0$) equation (3.10) determines the solution at a mesh point Q in the row $t = T + k$ by using the values of u along $t = T$ at the three adjacent points symmetrically placed with respect to Q. Now assume u is given for all t at two fixed values of x, say $x = a$ and $x = b$. These are typical boundary conditions for the diffusion equation as explained in Section 3.3. Then it follows that, starting from

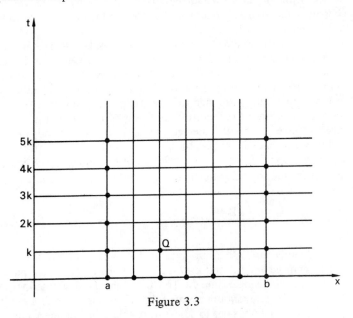

Figure 3.3

a knowledge of u at $t = 0$ in $a < x < b$, the values of u at the mesh points throughout the whole of the infinite rectangle $a < x < b, t > 0$, can be determined. In this process, it is assumed that the mesh is so chosen that the stability of the numerical procedure is assured. In this connection see Exercise 3.4 in which it is shown that it is not in general possible to solve equation (3.1) in the negative t-direction. Here again there is a difference from the behaviour of the wave equation and we shall return to this point in the next section.

3.5 Solution of the Diffusion Equation

We now consider the solution of the diffusion equation subject to suitable initial and boundary conditions. For the most part this consists of an application of the method of separation of variables as in Section 2.7 but in some cases where the domain of interest is infinite (Example (ii) of Section 3.5.2) or where a boundary condition is time-dependent (Section 3.5.3) the method may be inappropriate. We describe there different methods of solution. Since, historically, the diffusion of heat in a solid is, perhaps, the most important application we begin by solving a typical problem in this field.

3.5.1 Heat Flow in a Slab or Bar

We mentioned in Section 3.2.1 the circumstances in which the one-dimensional diffusion equation may represent the flow of heat in an infinite slab, with plane faces, say, at $x = 0$ and $x = l$, or a bar with insulated sides with ends at $x = 0$ and $x = l$. To be specific we shall speak of the bar and endeavour to model the behaviour of the poker to which we have previously referred.

For convenience we discuss first the cooling problem associated with the poker withdrawn from the fire. Therefore we consider a bar that has had one end, say $x = l$, in contact with a heat source at temperature θ_1 for a long time and its other end, at $x = 0$, maintained at temperature θ_0. Thus the temperature is assumed to have attained a steady state (see Section 3.3) and since the condition for this is $\partial\theta/\partial t = 0$ (that is, no change of temperature with time occurs at any point in the rod) the diffusion equation (3.1) reduces to the simple ordinary differential equation, $d^2\theta/dx^2 = 0$. The general solution of this equation is a linear function of x and to satisfy the boundary conditions at $x = 0$ and l we obtain

$$\theta = \theta_0 + \frac{(\theta_1 - \theta_0)x}{l}, \quad 0 < x < l. \tag{3.11}$$

This is the temperature distribution in the bar, when, at $t = 0$, the heat source is removed and the end $x = l$ is also exposed to, and subsequently maintained at, the temperature θ_0. The diffusion equation describes the way in which the temperature at any point in the bar varies, as a result of the heat flow set up, so as to adjust to the changed boundary condition. Mathematically, the problem is to solve (3.1) in $0 < x < l$, subject to (3.11) as the initial condition ($t = 0$), and we assume

$$\theta = \theta_0, \quad x = 0; \quad \theta = \theta_0, \quad x = l; \quad t > 0. \tag{3.12a, b}$$

Of course, in asserting that this mathematical problem describes the state of affairs in the poker we have ignored the heat loss from the curved surface and have assumed the ends to be at room temperature θ_0. To make the former assumption more plausible we could surround the poker with insulating material. The modification that allows a more realistic assumption on the boundary conditions, by applying Newton's Law of Cooling (Section 3.3), is described at the end of this section.

Since θ_0 is constant we can eliminate it from the boundary conditions by using $\theta' = \theta - \theta_0$ as a new dependent variable. The equation for θ' is still (3.1) but the initial condition (3.11) becomes

$$\theta' = (\theta_1 - \theta_0)x/l, \quad t = 0, \quad 0 < x < l, \tag{3.13}$$

and the boundary conditions (3.12) are simplified to $\theta' = 0$ at $x = 0$ and $x = l$ for $t > 0$.

As for the wave equation we assume a solution in the separated form,

$$\theta'(x, t) = X(x)T(t).$$

Then substitution into (3.1), with $u \equiv \theta'$, yields

$$\frac{1}{\alpha^2 T} \frac{dT}{dt} = \frac{1}{X} \frac{d^2 X}{dx^2}.$$

Again each side must equal the same negative constant since the left-hand side is independent of x and the right-hand one is independent of t. Hence,

$$\frac{d^2 X}{dx^2} + m^2 X = 0, \quad \frac{dT}{dt} + \alpha^2 m^2 T = 0,$$

where the constant has been chosen to be $-m^2$. The solution for X is as given in (2.47) but that for T is exponential. Thus

$$X = A \cos mx + B \sin mx, \quad T = C \exp(-\alpha^2 m^2 t), \tag{3.14}$$

where A, B and C are arbitrary constants. The solution for X is exactly the same as in the wave motion on the string, fixed at $x = 0$ and l, which was obtained in Section 2.7. The general solution satisfying the boundary conditions is thus found by superposition to be

$$\theta' = \sum_{n=1}^{\infty} C_n \sin\left(\frac{n\pi x}{l}\right) \exp\left(-\frac{\alpha^2 n^2 \pi^2 t}{l^2}\right).$$

The constants C_n are determined by the initial condition which, from (3.13), requires that

$$(\theta_1 - \theta_0)\frac{x}{l} = \sum_{n=1}^{\infty} C_n \sin\left(\frac{n\pi x}{l}\right), \quad 0 < x < l.$$

The C_n are thus the coefficients of the Fourier sine series for the function of x on the left-hand side. Accordingly,

$$C_n = \frac{2}{l} \int_0^l (\theta_1 - \theta_0) \frac{x}{l} \sin\left(\frac{n\pi x}{l}\right) dx = \frac{2(\theta_1 - \theta_0)(-1)^{n+1}}{n\pi},$$

and the solution is,

$$\theta'(x, t) = \frac{2(\theta_1 - \theta_0)}{\pi} \sum_{n=1}^{\infty} \frac{(-1)^{n+1}}{n} \sin\left(\frac{n\pi x}{l}\right) \exp\left(-\frac{\alpha^2 n^2 \pi^2 t}{l^2}\right).$$

$$(3.15)$$

We see from (3.15) that the solution does not oscillate in time. The time dependence is an exponential decay of each term in the series and the source of this behaviour, if we look back through the derivation, is the first order time derivative in the diffusion equation (in contradistinction to the wave equation where the second order time derivative leads to oscillatory solutions). This monotonic decay to an ultimate solution $\theta' = 0$ throughout the bar illustrates the manner in which heat conduction smoothes out temperature differences. Differences in quantities governed by the wave equation are not smoothed out but propagate throughout the medium. The diffusion equation thus expresses the Second Law of Thermodynamics in prescribing the direction in which the heat flow takes place. The time scale associated with the change-over from the initial steady state to the final one depends on the properties of the medium; from (3.15) it is proportional to l and inversely proportional to α^2. Thus long, good heat conductors are favourable to a rapid transition.

From the expression (3.15) we notice also that the solution does not exist for negative t since the series is then divergent. The contrast of this property of the equation with that of the wave equation, discussed in Section 2.4, reflects the distinction between the irreversible and reversible processes (in the thermodynamic sense) which the equations (3.1) and (2.1) describe, respectively. Here, energy of the system is not conserved but lost to the surroundings in the form of heat. Indeed, for positive t, and for times later than the initial moments of the heat flow (such that the argument of the first exponential, $\alpha^2 \pi^2 t/l^2$, is not much smaller than unity) the presence of the exponentially decaying factors renders the series much more rapidly convergent than those arising in the wave problems. Thus at $\alpha^2 t = l^2/\pi^2$, the exponential factor in the nth term is the n^2 power of $e^{-1} = 0.368$. After a relatively short time only the first term or two in the infinite series (3.15) makes a significant contribution and in Fig. 3.4 the temperature distribution in the bar, computed from the first two terms only of the series (corresponding to the first and second harmonics in Fig. 2.13) for $\alpha^2 \pi^2 t/l^2 = \frac{1}{2}$, 1, 2, is drawn and compared with the initial, linear temperature distribution.

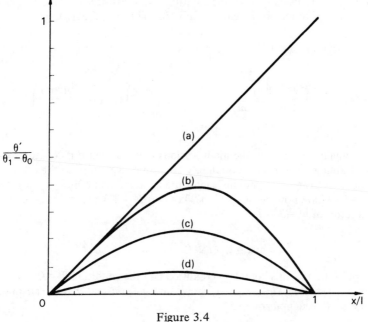

Figure 3.4

(a) $t = 0$, (b) $\alpha^2 \pi^2 t/l^2 = \frac{1}{2}$, (c) $\alpha^2 \pi^2 t/l^2 = 1$, (d) $\alpha^2 \pi^2 t/l^2 = 2$.

In order to discuss the converse problem of heating one end of the bar, initially at a uniform temperature θ_0, to a prescribed temperature θ_1 ('placing the poker in the fire'), we must solve equation (3.1) subject to

$$\theta(x, 0) = \theta_0, \quad 0 < x < l, \tag{3.16}$$

$$\theta(0, t) = \theta_0, \quad \theta(l, t) = \theta_1, \quad t > 0. \tag{3.17}$$

It is convenient to separate out from the full solution in this case the final steady state which corresponds to the solution of $d^2\theta/dx^2 = 0$ satisfying the boundary conditions (3.17). This, of course, is the initial state for our previous problem and is given by $\theta = \theta_s(x) = \theta_0 + (\theta_1 - \theta_0)x/l$, $0 < x < l$, from (3.11). We then write $\theta(x, t) = \theta_s(x) - \theta_t(x, t)$, where $\theta_t(x, t)$ also satisfies the diffusion equation, and on substitution into (3.16) and (3.17), we see that θ_t satisfies the subsidiary conditions,

$$\theta_t(x, 0) = (\theta_1 - \theta_0)(x/l), \quad 0 < x < l,$$
$$\theta_t(0, t) = \theta_t(l, t) = 0, \quad t > 0.$$

Hence the transient part of the solution, θ_t, is identical with the function θ' found above and the full solution is, from (3.15),

$$\theta(x, t) = \theta_0 + (\theta_1 - \theta_0)\frac{x}{l}$$

$$-\frac{2(\theta_1 - \theta_0)}{\pi} \sum_{n=1}^{\infty} \frac{(-1)^{n+1}}{n} \sin\left(\frac{n\pi x}{l}\right) \exp\left(-\frac{\alpha^2 n^2 \pi^2 t}{l^2}\right).$$

$$(3.18)$$

Finally we consider the modifications that are needed if Newton's Law of Cooling is taken as the boundary condition. We return to the cooling problem treated first and instead of the boundary condition (3.12b) at $x = l$ we have now, using (3.8), $\partial\theta/\partial x = -h(\theta - \theta_0)$, at $x = l$, $t > 0$, or, in terms of $\theta' = \theta - \theta_0$,

$$\frac{\partial \theta'}{\partial x} = -h\theta', \quad \text{at } x = l, \quad t > 0.$$

Application of this boundary condition and the condition $\theta' = 0$ at $x = 0$ on the function $X(x)$ in (3.14) shows that the constant m now satisfies the transcendental equation,

$$h \tan ml = -m.$$

Like the earlier cases there are infinitely many solutions of this equation, as can be seen graphically from the intersection of the curves $y = \tan ml$ with $y = -m/h$. If we denote the sequence of positive solutions in ascending order of magnitude by $m = m_n, n = 1, 2, 3, \ldots$ then a solution of the diffusion equation, which satisfies also the required boundary conditions at $x = 0$ and $x = l$, is

$$\theta' = C_n \exp(-\alpha^2 m_n^2 t) \sin(m_n x), \tag{3.19}$$

for each positive integral value of n and arbitrary C_n. Because of linearity and homogeneity of the boundary conditions, any sum of solutions of the form (3.19) has the same properties and, therefore, we can find a complete solution of our problem if we can find such a sum (that is, determine the C_n) in such a way that the initial condition is also satisfied. To do this we require

$$(\theta_1 - \theta_0)\frac{x}{l} = \sum_{n=1}^{\infty} C_n \sin(m_n x).$$

The series is not now a Fourier sine series because the m_n are no longer linearly dependent on n. However, although it is beyond the scope of this book to pursue the details, it is true that the functions $\sin(m_n x)$ are orthogonal on the interval $0 < x < l$. That is to say

$$\int_0^l \sin(m_r x) \sin(m_s x)\, dx = 0, \quad \text{if } r \neq s,$$

and the integral clearly does not vanish when $r = s$. On using this property and assuming the series can be integrated term-by-term, which can be justified, the coefficients C_n are seen to be

$$C_n = \frac{(\theta_1 - \theta_0) \int_0^l x \sin(m_n x)\, dx}{l \int_0^l \sin^2(m_n x)\, dx}.$$

3.5.2 Current Flow in a Cable

As a second example we consider the flow of electric current in a cable under conditions for which the diffusion equation applies (Section 3.2.2).

Example (i). First suppose the cable extends from $x = 0$ to $x = l$ and steady state conditions exist in which the potential at $x = 0$ is V_0 and at $x = l$ is V_1 ($< V_0$) when at $t = 0$ the end $x = l$ is suddenly earthed. The problem is to calculate the current in the cable given that the resistance and capacitance per unit length are R and C respectively. In this model of the cable we recall that the conductance from cable sheath to earth and the inductance are neglected.

We first note that it is the current I that is required although the boundary conditions are imposed on the potential V. However, a knowledge of V carries with it a knowledge of I from (3.5) and so we first find V. The boundary conditions on V are

$$V = V_0 \text{ at } x = 0, \quad V = 0 \text{ at } x = l, \quad t > 0. \tag{3.20}$$

The initial condition is the steady state from V_0 at $x = 0$ to V_1 at $x = l$ before the connection to earth. Hence it is again a linear function,

$$V = V_0 + (V_1 - V_0)\frac{x}{l}, \quad t = 0, \quad 0 < x < l. \tag{3.21}$$

As in the previous example we write $V = V_s(x) + V_t(x, t)$, where $V_s(x)$ is the final steady state and the transient part $V_t(x, t)$ describes the transition from the initial state (3.21). (Since the final potential in this case will be less than the initial one for any $x > 0$, the plus sign attached to V_t ensures that V_t is positive.) $V_s(x)$ is the time-independent solution of the diffusion equation satisfying (3.20),

$$V_s = V_0\left(1 - \frac{x}{l}\right),$$

and $V_t(x, t)$ satisfies the diffusion equation subject to, $V_t(0, t) = 0$, $V_t(l, t) = 0$, $t > 0$: $V_t(x, 0) = V_1 x/l$, $0 < x < l$. These conditions on V_t

are essentially the same as the conditions on θ_t in the heating problem for the bar discussed in Section 3.5.1. Hence we can use the solution obtained in that case, (3.15), to write down the solution for V_t. The result for V is

$$V = V_0\left(1 - \frac{x}{l}\right) + \frac{2V_1}{\pi} \sum_{n=1}^{\infty} \frac{(-1)^{n+1}}{n} \sin\left(\frac{n\pi x}{l}\right) \exp\left(-\frac{\alpha^2 n^2 \pi^2 t}{l^2}\right),$$

(3.22)

where $\alpha^2 = (RC)^{-1}$. The current is given by

$$I = -\frac{1}{R}\frac{\partial V}{\partial x} = \frac{V_0}{Rl} - \frac{2V_1}{Rl} \sum_{n=1}^{\infty} (-1)^{n+1} \cos\left(\frac{n\pi x}{l}\right) \exp\left(-\frac{\alpha^2 n^2 \pi^2 t}{l^2}\right).$$

For a given value of t the series in I converges less rapidly than that in V. Notice also that a good conductor (low R, high α^2) leads to a transient of relatively short duration.

Example (ii).   The example just considered, concerning the current flow in a cable, suggests another problem that could be posed. Suppose the cable to be so long that we can ignore the effects of one end which is therefore taken to be at $x = \infty$, whilst at the other end, $x = 0$, a constant potential V_0 is maintained for $t > 0$. Initially the whole cable is assumed to be at zero potential. Thus, a solution of the diffusion equation is sought which satisfies the initial condition,

$$V = 0, \quad 0 < x < \infty, \quad t = 0,$$

(3.23)

and the single boundary condition,

$$V = V_0, \quad x = 0, \quad t > 0.$$

(3.24)

The student will soon convince himself that an attempt to tackle the problem by using the method of separation of variables fails, essentially because of the infinite x-domain. However, the problem provides a good illustration of a different approach, motivated by physical arguments and widely used by physicists and engineers.

The crucial difference between this example and the last one is that, in considering an infinite cable, we have removed a length scale, the length l of the cable in Example (i), from the whole problem. Hence, whatever its actual value, the dimensionless form of the solution, V/V_0, must be expressible in the form,

$$V/V_0 = f(\alpha^2, x, t),$$

(3.25)

where f is also a dimensionless expression for all the remaining parameters and variables in the problem. This follows from the fact that the two sides of an identity must have the same dimensions. Now it was pointed out in Section 3.1 that α^2 has dimensions of $(\text{length})^2/\text{time}$ and hence f in (3.25) must depend on α^2, x and t in the form $\alpha^2 t/x^2$, the only dimensionless combination that can be found. Thus,

$$V/V_0 = f(\eta), \quad \eta = \alpha^2 t/x^2, \tag{3.26}$$

and we may use this as the assumed form of the solution of the diffusion equation subject to the conditions (3.23) and (3.24).

Notice that with the further length scale, l, available in Example (i), V/V_0 can be expressed in terms of two dimensionless variables, x/l and $\alpha^2 t/l^2$, which are precisely the combinations that appear in the solution (3.22) obtained by separation of variables.

Partial derivatives with respect to x and t transform to η-derivatives, on using (3.26), according to the formulae,

$$\frac{\partial}{\partial t} = \frac{1}{\alpha^2 x^2} \frac{d}{d\eta}, \quad \frac{\partial}{\partial x} = -\frac{2t}{\alpha^2 x^3} \frac{d}{d\eta}.$$

Thus, the diffusion equation (3.1) becomes,

$$\frac{d^2 V}{d\eta^2} + \left(\frac{3}{2\eta} - \frac{1}{4\eta^2} \right) \frac{dV}{d\eta} = 0, \tag{3.27}$$

which is a linear, first-order, ordinary differential equation for $dV/d\eta$. The integrating factor is,

$$\exp\left\{ \int \left(\frac{3}{2\eta} - \frac{1}{4\eta^2} \right) d\eta \right\} = \eta^{3/2} \exp(1/4\eta).$$

Hence the equation can be integrated twice to give,

$$V = K_1 \int_\infty^\eta \eta^{-3/2} \exp(-1/4\eta) \, d\eta + K_2,$$

where K_1 and K_2 are constants of integration. The integral can be simplified by introducing a new variable of integration, σ, defined by, $\sigma^2 = 1/4\eta$. We then obtain,

$$V = K_1 \int_0^{x/2\alpha\sqrt{t}} \exp(-\sigma^2) \, d\sigma + K_2, \tag{3.28}$$

where a constant has been absorbed into K_1. Actually, the integral,

$$\frac{2}{\sqrt{\pi}} \int_0^y \exp(-\sigma^2) \, d\sigma = \operatorname{erf}(y),$$

where 'erf' stands for 'error function', is well-known in mathematical literature and is given in mathematical tables.

The upper limit of the integral in (3.28) vanishes at $x = 0$, $t > 0$ and, hence, the boundary condition (3.24) yields $K_2 = V_0$. Further, this upper limit is unbounded at $t = 0$ for $x > 0$ and since $\mathrm{erf}(y) \to 1$ as $y \to \infty$, the initial condition (3.23) yields $K_1 = -2V_0/\sqrt{\pi}$. Thus the two conditions (3.23) and (3.24) are sufficient to obtain a unique solution of the problem, and it may be written as

$$V/V_0 = 1 - \mathrm{erf}(x/2\alpha\sqrt{t}).$$

This expression is instructive in describing how the potential diffuses along the cable. At a given instant of time there are large enough values of x for t to be effectively zero; on the other hand at any given x, V eventually approaches V_0. One desirable requirement in actual cables is that this diffusion process should take place in a sufficiently short time so as to maintain the separation of successive pulses. This means that α should be large, or from Section 3.2.2, the resistance should be sufficiently low; but the thicker cable that this entails is limited by cost requirements and other means are used in practice.

The characteristic feature of the transformation of the type (3.26) used in this method of obtaining solutions of partial differential equations is a reduction in the number of independent variables, from two (x, t) to one (η) in this case. It is an example of a *similarity* transformation in mathematics. The name derives from the fact that for arbitrary fixed times, $t = t_i$, say, all the potential profiles $V = V(x, t_i)$ along the cable can be plotted using the single function $V = V_0[1 - \mathrm{erf}(\eta)]$; in that sense the profiles are 'similar'.

3.5.3 Shear Waves in a Viscous Fluid

As a final example we consider a somewhat different type of problem that can arise in connection with the diffusion equation. With time-independent boundary conditions, we have seen that the solutions are not oscillatory in character. However, oscillations can be 'forced' on the solution through the boundary conditions. We illustrate this possibility by returning to the situation, described in Section 3.2.3, of the motion of an infinite volume of viscous, incompressible fluid bounded by a plane wall. Under certain conditions the velocity of the fluid has a single non-vanishing component, parallel to the wall, which satisfies the diffusion equation subject to the boundary conditions, discussed in Section 3.3, of which one is determined by the velocity in its own plane of the wall itself.

To be specific we suppose the velocity of the infinite wall, which lies in the plane $z = 0$, is

$$v_w = U \cos nt, \tag{3.29}$$

where U and n are constant, and this induces the fluid motion through the no-slip boundary condition which is $v = v_w$ on $z = 0$ for all t.

Of course, in practice the wall would have started up from rest and in the process of acquiring the form (3.29) a transient fluid motion would have been set up which we assume has decayed away. In other words, we assume that sufficient time has elapsed since the start of the motion that the fluid velocity is determined *only* through the boundary conditions and differential equation and has become independent of the actual initial condition. This means that we need to calculate that part of the solution which in the previous examples we called u_s, the final steady state. Now, however, this solution will be time-dependent and we assume on physical grounds that the dependence on t follows that of the wall in being periodic with frequency n. We can think of this condition on the solution as that which replaces the 'lost' initial condition in the present problem.

It is again convenient as in Section 2.5 to use complex variables, with real parts representing the physical velocity. Thus, instead of (3.29) we write the wall velocity in the form,

$$v_w = \mathrm{Re}[U\, e^{int}], \quad U \text{ real},$$

and the fluid velocity,

$$v = \mathrm{Re}[f(z)\, e^{int}], \tag{3.30}$$

where $f(z)$ is a complex-valued function which is to be found. The boundary condition on the wall then yields $f(0) = U$ and the remaining boundary condition is that $f(z) \to 0$ as $z \to \infty$ from the physical condition that the disturbance caused by the wall creates no effect in the fluid infinitely far from it (see Section 3.3). On substituting (3.30) into the diffusion equation (3.1) with $\alpha^2 = v$, the viscosity, we find that f satisfies,

$$\frac{d^2 f}{dz^2} - \frac{inf}{v} = 0.$$

This is an ordinary differential equation with a constant (complex) coefficient and standard methods yield the general solution,

$$f(z) = A \exp[(1+i)(n/2v)^{1/2} z] + B \exp[-(1+i)(n/2v)^{1/2} z],$$

where A and B are constants to be determined from the boundary conditions. Since f must vanish for large z, $A = 0$ and the boundary condition on $z = 0$ then yields $B = U$. Thus the solution is,

$$v = \mathrm{Re}\{U \exp[-(1+i)(n/2v)^{1/2} z + int]\}$$

$$= U \exp[-z(n/2v)^{1/2}] \cos[nt - (n/2v)^{1/2} z]. \tag{3.31}$$

At fixed z the velocity oscillates at the same frequency as the wall but with an amplitude which is smaller by a factor $\exp[-z(n/2v)^{1/2}]$ and with a phase that lags behind that of the wall by an amount $z\sqrt{(n/2v)}$. At a distance $\sqrt{(2v/n)}$ from the wall the velocity amplitude falls by a factor $1/e$. This distance is sometimes called the *penetration depth*. From what we have seen in Chapter 2 we can regard the solution (3.31) as representing a transverse, harmonic wave propagating normally from the wall into the fluid with an amplitude which is decreasing monotonically with distance. The phase velocity of the waves is $\sqrt{(2vn)}$. Such waves are known as *shear waves* since they are associated with velocity gradients in the flow.

The frictional force F_w acting on unit area of wall can be calculated from the solution (3.31). The only component is parallel to the wall and is given by $\rho v \partial v / \partial z$, evaluated at the wall $z = 0$, where ρ is the density. Thus, we obtain from (3.31),

$$F_w = U\rho\sqrt{(nv/2)}(-\cos nt + \sin nt) = -U\rho\sqrt{(nv)}\cos(nt + \pi/4).$$

(3.32)

Hence this force, the *skin friction*, has also a phase difference compared with the motion of the wall. We can use the result to find the energy dissipated per unit time per unit area of wall by means of the fluid's viscosity. This quantity is equal to the average value of the rate of working of the frictional force,

$$-\frac{\displaystyle\int_{t_0}^{t_0+2\pi/n} F_w v_w \, dt}{2\pi/n} = \frac{\displaystyle\int_{t_0}^{t_0+2\pi/n} U^2\rho\sqrt{(nv)}\cos nt \cos(nt + \pi/4) \, dt}{2\pi/n}$$

$$= \frac{\rho U^2}{2}\sqrt{\left(\frac{nv}{2}\right)},$$

where t_0 is an arbitrary fixed time, and is seen to be proportional to the square root of both the frequency and viscosity.

The problem discussed in this section also has application in determining the thermal response of the earth's surface layers to the annual and diurnal heating and cooling of the earth's surface by the sun (see Exercises 3.14 and 3.15). Other exercises involve different wall velocities or different geometries.

EXERCISES

1.   Suppose the volume $\delta\tau$ in Fig. 3.1 contains a mixture of two different gases. If the concentration c of one gas in the mixture is sufficiently low, the rate per unit area at which it flows across a plane is proportional to the normal derivative of the concentration on the plane (cf. equation (3.2)). By considering the conservation of mass of the dilute gas in $\delta\tau$ derive the equation, $\partial c/\partial t = D\nabla^2 c$, governing the manner of its diffusion in the mixture. (D is the coefficient of diffusion of the dilute gas in the mixture.)

2.   The temperature profile between points $x = 0$, where the temperature is zero, and $x = 2h$, where it is Θ, is initially linear. Using the difference equation (3.10) show that the temperature at $x = h$ decreases

when the temperature at $x = 2h$ is reduced and increases when it is raised.

3. At some instant there is a parabolic temperature profile connecting temperatures of 0 at $x = 0$ and Θ at $x = 2h$. After one time increment use (3.10) to show that there is a rise or fall of temperature at $x = h$ according as the curvature of the initial profile is positive or negative.

4. Suppose the temperature distribution is given for $t = 0$ in $-\infty < x < \infty$. Show from the difference equation (3.10) that the solution can be obtained for the mesh points in the (x, t) plane satisfying $t > 0$, but not for $t < 0$.

5. Solve $\partial u/\partial t = \alpha^2 \, \partial^2 u/\partial x^2$ subject to the conditions, $u(x, 0) = x^2$, $0 < x < l; u(0, t) = 0, t > 0$ and $\partial u/\partial x = 0, x = l, t > 0$, and suggest a suitable physical problem to which the model might apply.

6. A metal plate with plane faces at $x = 0$ and $x = l$ extends to infinity in the y and z directions. Its initial temperature distribution is $\theta_0 \sin^2(\pi x/2l)$. Find the temperature distribution in the plate if the temperatures of the faces are suddenly changed to θ_1.

7. Repeat Exercise 6 when the face $x = 0$ is heat-insulated and the temperature θ_1 is imposed on the face $x = l$ at $t = 0$.

8. Consider a cable of length l under conditions in which the diffusion equation applies (see Section 3.2.2). Initially the potential is zero everywhere, when the end $x = 0$ is raised to a constant potential V. Obtain the current at any point in the cable and discuss the validity of the term-by-term differentiation of the series (cf. Section 1.2).

9. A uniform rod of length l has a temperature distribution $x(l - x)$. At time $t = 0$ the ends of the rod are heat-insulated. Neglecting the heat losses over the curved surface of the rod find the temperature distribution for $t > 0$.

10. Repeat Example (ii), Section 3.5.2, for the case where the cable in $0 < x < \infty$ is initially at a potential V_0 and the end $x = 0$ is then earthed for $t > 0$. Discuss the behaviour of the current.

11. Find the velocity distribution in a viscous fluid bounded by an infinite plane wall, $z = 0$, when the wall is given a constant velocity U in its own plane for $t > 0$. The fluid and wall are initially at rest. Interpret the solution and calculate the frictional force acting on unit area of wall. Deduce that the layer of fluid next to the wall outside which the velocity is less than a fixed fraction of U has a thickness proportional to $\sqrt{(vt)}$. (Hint. Use the method of Example (ii), Section 3.5.2.)

12. Find the velocity distribution in a viscous fluid when an infinite plane wall, $z = 0$, oscillates sinusoidally in its own plane, as in Section 3.5.3, but the fluid occupying $z > 0$ is bounded by another stationary plane wall at $z = l$. Find also the frictional force acting on unit area of the wall at $z = 0$.

13.   Repeat Exercise 12 when both walls at $z = 0$ and $z = l$ oscillate in their own planes with velocity $v \cos nt$.

14.   Suppose locally the surface of the earth is considered plane. The periodic temperature distribution, $\theta_0 + \theta_1 \cos nt$, represents the daily variation of the heating and cooling of the surface due to the sun's rays. Find an expression for the penetration depth in terms of the diffusion coefficient and obtain the maximum fluctuation of temperature at depth x.

15.   In the preceding problem assume that the surface layer of thickness l is effectively separated from the material below by a surface at $x = l$ across which no heat can flow. Find the temperature variation at this boundary. (The positive direction of x is downwards from the earth's surface.)

CHAPTER 4

Laplace's Equation

4.1 Introduction

In this chapter we consider the third of the standard partial differential equations of second order,

$$\frac{\partial^2 u}{\partial x^2} + \frac{\partial^2 u}{\partial y^2} = 0. \tag{4.1}$$

We choose y as the second independent variable because in usual applications both independent variables refer to space rather than time. Equation (4.1) is known as the two-dimensional Laplace's equation and is the simplest non-trivial example of the class of partial differential equations called elliptic (Section 1.3).

We have met already, in Chapter 3, a degenerate form of (4.1) as the equation governing suitable initial and final states in the solution of the diffusion equation. Thus, for example, in the heating problem for the poker (assuming a one-dimensional model), described in Section 3.5.1, the final steady state (that is, for large time) is the linear function of x that yields the maintained temperatures at the two ends. The temperature satisfies the time-independent diffusion equation for that case, $d^2u/dx^2 = 0$, and we may adopt a similar view here with regard to (4.1) except that the number of space dimensions on which the variable u can depend is increased to two. Thus equation (4.1) is associated with the study of steady state phenomena and although the one-dimensional version of it is too simplified for many purposes it does provide a clue to the kind of behaviour we might expect. For instance, at any point P on the poker (in the final state) the temperature is the mean of the temperatures at two points equidistant from, and on either side of P; in that sense the equation represents an averaging process with no maximum or minimum temperature occurring in the interior of the poker. Furthermore, the boundary conditions are applied over the entire boundary, which consists in this simple case of the two points at which the temperature is prescribed. We shall see that these properties are typical of the more general equation (4.1).

Solutions of Laplace's equation are called harmonic functions and the term 'harmonic equation' is also used for it. Because of its close connection with potential theory (see Section 4.2.2) equation (4.1) is sometimes referred to as the potential equation although this is not strictly accurate since a potential function (the gradient of which represents some vector field) may not satisfy (4.1), nor need the function u in (4.1) represent a potential in the physical sense.

The pattern of this chapter follows that of Chapters 2 and 3. Situations where Laplace's equation serves as a model are described in Section 4.2 and the associated boundary conditions are discussed in Section 4.3. Questions on the determinacy of solutions and some of the chief properties, including the connection with analytic functions of a complex variable, are considered in Section 4.4. Lastly, in Section 4.5, the method of separation of variables is applied to obtain solutions in Cartesian and plane polar coordinates.

4.2 Applications of Laplace's Equation

4.2.1 Steady Flow of Heat in a Solid

In Section 3.2.1 the general equation, (3.4), governing the non-steady flow of heat by conduction in a solid, in the full three-dimensional case, was derived. This equation becomes the two-dimensional Laplace's equation (4.1) under the conditions,

  (i) the temperature is independent of time,
  (ii) the temperature is independent of the z-coordinate.

The first condition implies that the flow of heat has been going on so long that a steady state has been reached; a graph of the temperature distribution along any line through the solid looks the same at different times. The second condition means that the temperature distribution is the same in all planes parallel to the (x, y) plane and, thus, restricts the application of equation (4.1) to bodies such as infinite cylinders or finite cylinders (or laminae) with heat insulated ends, having generators parallel to the z-axis. Subject to (i) and (ii), then, the temperature within the body satisfies (4.1).

4.2.2 Irrotational, Solenoidal Vector Fields

Frequently, in the study of continuous physical systems the object of the investigator is to obtain the value of some vector quantity at any instant throughout the volume of interest, τ, say. Examples of such vector fields, (that is, vector functions of position), some of which we have met already in earlier chapters, are the electric and magnetic fields in the theory of electromagnetism, the gravitational field, or force per unit mass on a test particle, in the theory of gravitational attraction and the velocity field in a moving fluid. Thus the solution of a problem in these areas is the determination of a vector quantity, or, in general, three scalar quantities, at each point of τ. Under certain conditions, however, the problem can be reduced to the much simpler one of obtaining a single scalar function at each point, a scalar field, which satisfies Laplace's equation, $\nabla^2 u = 0$, where ∇^2 is the Laplacian operator $\partial^2/\partial x^2 + \partial^2/\partial y^2 + \partial^2/\partial z^2$ in Cartesian coordinates.

There are two conditions to be satisfied by the vector field, **F**, say, if this simplification is to be possible. In the first place, **F** must satisfy the equation curl **F** = 0 everywhere in τ. If this is so then an application of Stokes theorem round any closed circuit C in τ, which encloses a surface S lying entirely in τ, shows that

$$\int_C \mathbf{F}\,.\,ds = \int_S \text{curl } \mathbf{F}\,.\,d\mathbf{S} = 0, \tag{4.2}$$

where ds is an element of C and $d\mathbf{S}$ an element of S, the directions of these elementary vectors being along the tangent to C and the normal of S, respectively, and their positive senses related by the right-hand screw rule. The quantity $\int_C \mathbf{F}\,.\,ds$ is called the *circulation* of **F** round C and vectors which have zero circulation about any closed curve C in τ are called *irrotational*.

Now if **F** is irrotational the integral between any two points P_1 and P_2,

$$\int_{P_1}^{P_2} \mathbf{F}\,.\,ds, \tag{4.3}$$

is unambiguous because by (4.2) the line integral along an arbitrary closed path, passing through P_1 and P_2, vanishes and hence $\int_{P_1}^{P_2} \mathbf{F}\,.\,ds$ has the same value along either branch going from P_1 to P_2 (see Fig. 4.1). In other words (4.3) is independent of the path. It follows that $\int_{P_1}^{P_2} \mathbf{F}\,.\,ds$ can be written in the form $u(P_2) - u(P_1)$ where u is a scalar field.

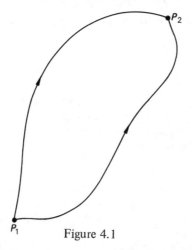

Figure 4.1

Now suppose P_1 is the point (x, y, z) and P_2 the point $(x + \delta x, y, z)$. Then,

$$u(x + \delta x, y, z) - u(x, y, z) = \int_{x}^{x+\delta x} F_x\,dx,$$

where $\mathbf{F} = (F_x, F_y, F_z)$. On dividing by δx and letting δx approach zero, we obtain in the limit, $F_x = \partial u/\partial x$. Similarly, by considering elementary paths in the y and z directions, we obtain, $F_y = \partial u/\partial y$, $F_z = \partial u/\partial z$. We deduce that if $\mathbf{F}$ is irrotational then there exists a scalar field u such that

$$\mathbf{F} = \text{grad } u = \frac{\partial u}{\partial x}\mathbf{i} + \frac{\partial u}{\partial y}\mathbf{j} + \frac{\partial u}{\partial z}\mathbf{k}, \tag{4.4}$$

where $\mathbf{i}, \mathbf{j}$ and $\mathbf{k}$ are unit vectors in the x, y and z coordinate directions. In some instances it is conventional to include a minus sign and define u by $\mathbf{F} = -\text{grad } u$. The scalar field u is called the *potential* of $\mathbf{F}$.

Secondly, $\mathbf{F}$ may be further restricted by the equation,

$$\text{div } \mathbf{F} = \frac{\partial F_x}{\partial x} + \frac{\partial F_y}{\partial y} + \frac{\partial F_z}{\partial z} = 0. \tag{4.5}$$

$\mathbf{F}$ is then said to be a *solenoidal* vector. On substituting equation (4.4) into (4.5) we obtain

$$\text{divgrad } u = \nabla^2 u = \frac{\partial^2 u}{\partial x^2} + \frac{\partial^2 u}{\partial y^2} + \frac{\partial^2 u}{\partial z^2} = 0, \tag{4.6}$$

as the equation to be satisfied by u. When u is found from this equation, for example by a method described in Section 4.5, then the vector field $\mathbf{F}$ is determined from (4.4). In two dimensions, of course, equation (4.6) becomes (4.1).

As one example from electromagnetic theory we may take the Maxwell equations in MKS units

$$\text{curl } \mathbf{E} = -\mu \frac{\partial \mathbf{H}}{\partial t}, \quad \text{div } \mathbf{E} = \frac{\sigma}{\epsilon},$$

where $\mathbf{E}$ and $\mathbf{H}$ are the electric and magnetic intensities, σ is the charge density and ϵ and μ are the dielectric constant and permeability of the medium. In steady conditions in regions where there is no distribution of charge these equations reduce to

$$\text{curl } \mathbf{E} = 0, \quad \text{div } \mathbf{E} = 0, \tag{4.7a, b}$$

and $\mathbf{E}$ is both irrotational and solenoidal. Consequently it suffices to solve $\nabla^2 u = 0$, where $\mathbf{E} = -\text{grad } u$ and u is the electrostatic potential. Equation (4.7b) gave rise to the adjective 'solenoidal' since it is characteristic of conditions inside a solenoid.

A second example arises in the theory of gravitational attraction. Newton's inverse square law states that the force on a particle of unit mass at a point P due to a particle of mass m at a point Q is given by

$$\mathbf{F} = -\frac{Km\mathbf{r}}{r^3}, \tag{4.8}$$

where $\mathbf{r}$ is the displacement vector $\overrightarrow{QP}$, r its magnitude and K is a constant. For example, if the Cartesian coordinates of Q and P referred to some origin 0 are (x_1, y_1, z_1) and (x, y, z), then $\mathbf{r} = (x - x_1)\mathbf{i} + (y - y_1)\mathbf{j} + (z - z_1)\mathbf{k}$. We can readily verify (Exercise 4.1) that $\mathbf{F}$ is both irrotational and solenoidal for $r \neq 0$. Hence there exists a scalar potential $u(x, y, z)$ such that $\mathbf{F} = -\nabla u$ and $\nabla^2 u = 0$. This analysis is for a point mass at Q but it can be extended by superposition to include the case of matter distributed continuously throughout a volume τ. Then the contribution to the force at P (assumed outside τ) due to each volume element $\delta\tau_i$ of τ is of the form (4.8), provided $\mathbf{r}$ is interpreted as the vector joining the element to P and m as the mass of the element, $\rho\delta\tau_i$, where ρ is the density; the resultant force could be calculated by integration over τ. The potential at P due to the mass distribution in τ again satisfies $\nabla^2 u = 0$ and the gravitational force per unit mass is $-\text{grad } u$.

Finally, we return to the motion of a fluid such as has been previously considered in Section 3.2.3. If the fluid has a velocity field $\mathbf{v}$ and is assumed incompressible, equation (3.6) holds, or, in vector form,

$$\text{div } \mathbf{v} = 0. \tag{4.9}$$

The vector form of the momentum equations (3.7) is

$$\frac{\partial \mathbf{v}}{\partial t} + \mathbf{v} \cdot \text{grad } \mathbf{v} = -\frac{1}{\rho} \text{grad } p + \nu\nabla^2\mathbf{v}, \tag{4.10}$$

where $\mathbf{v} \cdot \text{grad } \mathbf{v} = (\mathbf{v} \cdot \text{grad } v_1, \mathbf{v} \cdot \text{grad } v_2, \mathbf{v} \cdot \text{grad } v_3)$ and $\nabla^2\mathbf{v} = (\nabla^2 v_1, \nabla^2 v_2, \nabla^2 v_3)$ in Cartesian coordinates. We now use the vector identity,

$$(\mathbf{v} \cdot \text{grad})\mathbf{v} = \text{grad}\left(\frac{v^2}{2}\right) - \mathbf{v} \wedge \text{curl } \mathbf{v},$$

and write curl $\mathbf{v} = \boldsymbol{\omega}$, the vorticity. Then equation (4.10) becomes

$$\frac{\partial \mathbf{v}}{\partial t} + \text{grad}\left(\frac{v^2}{2}\right) - \mathbf{v} \wedge \boldsymbol{\omega} = -\frac{1}{\rho} \text{grad } p + \nu\nabla^2\mathbf{v}. \tag{4.11}$$

On taking the curl of this equation we obtain

$$\frac{\partial \boldsymbol{\omega}}{\partial t} - \text{curl}(\mathbf{v} \wedge \boldsymbol{\omega}) = \nu\nabla^2\boldsymbol{\omega},$$

since curl grad $u \equiv 0$ for all u. One possible class of solutions of this equation is that for which $\boldsymbol{\omega} = \text{curl } \mathbf{v} \equiv 0$. For these so-called irrotational motions there exists a velocity potential u which from (4.9) satisfies

$$\nabla^2 u = 0. \tag{4.12}$$

However, regions of flow in which the fluid's viscosity plays an important role are nearly always rotational (so that $\boldsymbol{\omega} \neq 0$). Consequently, equation (4.12) is mainly used in regions where viscosity effects may be neglected, for example, away from the neighbourhood of solid boundaries. It is usual in hydrodynamics to define u such that $\mathbf{v} = +\text{grad } u$.

4.2.3 The Stream Function in Hydrodynamics

In the last example we have seen that in irrotational motion of an incompressible fluid the velocity potential u satisfies $\partial^2 u/\partial x^2 + \partial^2 u/\partial y^2 = 0$ in two dimensions. There is an alternative way of formulating the motion in terms of a scalar field which is sometimes preferable. Thus, instead of postulating the existence of a potential from the equation curl $\mathbf{v} = 0$, we can say that div $\mathbf{v} = \partial v_1/\partial x + \partial v_2/\partial y = 0$, is satisfied if there exists a scalar field U such that,

$$v_1 = \frac{\partial U}{\partial y}, \quad v_2 = -\frac{\partial U}{\partial x}. \tag{4.13}$$

On substituting these expressions into curl $\mathbf{v} = 0$ we obtain,

$$\frac{\partial^2 U}{\partial x^2} + \frac{\partial^2 U}{\partial y^2} = 0.$$

The advantages of using the *stream function* U stem from the fact that it is constant along curves, called streamlines, which at each point have the same direction as the local velocity vector.

Actually, the fact that div curl $\mathbf{U} \equiv 0$ for any differentiable vector field $\mathbf{U}$ suggests that a solenoidal vector may be expressed as the curl of a vector $\mathbf{U}$ (the vector potential). With the restriction to two dimensions, the scalar function U defined by (4.13) corresponds to the third component of $\mathbf{U}$ and, in this case, is the only one that plays a significant role.

4.3 Boundary Conditions

Since time is not an independent variable in equation (4.1) it is no longer appropriate to distinguish between initial and boundary conditions. However, the region of interest in a problem is now a two-dimensional one and, as noted in Section 1.4, conditions must be applied all round the boundary, in general, even if one is infinitely far away. The form in which the information on the boundary is given depends on the physical problem under discussion. Thus data comprising prescribed values of u, or its normal derivative, or some relation between them are all possible and different forms may apply for different parts of the boundary.

On referring to the examples described in the previous section, in the electrostatic and gravitational applications it is usually the value of the potential itself which is prescribed on the boundary, whilst in the hydrodynamic case it is often the normal derivative of the velocity potential that is given in order that the kinematic condition of zero relative normal velocity at a solid boundary be satisfied. The normal velocity component at a point on the boundary surface is

$$\mathbf{v}.\mathbf{n} = \operatorname{grad} u.\mathbf{n} = \frac{du}{dn}, \tag{4.14}$$

where $\mathbf{n}$ is the unit normal, directed towards the solid.

Sometimes Laplace's equation is to be solved in a volume of liquid which may have a free surface; that is a part of its boundary may be in contact with another fluid such as air. In that event the boundary condition that is appropriate is the continuity of stress across the interface of the two media, or, with viscosity and surface tension effects neglected, the boundary condition reduces simply to the continuity of pressure across the interface. Two new complications now arise:

(i) the condition on pressure must be transformed into one on the velocity potential,

(ii) the location of the interface is usually to be found as part of the problem.

As an example of how these difficulties may be circumvented we return to the case of water waves, of which we met one type, with negligible vertical acceleration, in Section 2.2.4. If a conservative body force of potential G is included and the viscous term omitted, the momentum equation for an incompressible fluid (4.11), becomes

$$\operatorname{grad}\left(\frac{v^2}{2} + \frac{p}{\rho} + G\right) = -\frac{\partial \mathbf{v}}{\partial t} = -\operatorname{grad}\left(\frac{\partial u}{\partial t}\right),$$

since we assume $\mathbf{v} = \operatorname{grad} u$ and therefore $\boldsymbol{\omega} = \operatorname{curl} \mathbf{v} = 0$. This equation can be integrated with respect to the space coordinates to give

$$\frac{v^2}{2} + \frac{p}{\rho} + G = -\frac{\partial u}{\partial t}. \tag{4.15}$$

The arbitrary function of time from the integration has been absorbed into u without loss of generality since fluid velocities, being space derivatives, are unaffected in so doing. As in Section 2.2.4 we consider the disturbance to the water to be so small that the equation can be linearized. Thus the term $v^2/2$ in (4.15) can be neglected and the relation between p and u becomes, $p = -\rho(gz + \partial u/\partial t)$, where we have written $G = gz$ to denote the gravitational potential. The boundary condition of continuity of pressure across the free surface $z = Z(x, t)$ becomes the condition,

$$\left(\frac{\partial u}{\partial t}\right)_{\delta = Z} = -gZ, \tag{4.16}$$

for u and Z where the uniform air pressure has been absorbed into u. For small Z, the derivative on the left-hand side may be evaluated at $z = 0$ to leading order (cf. Section 2.5.3) so that (4.16) becomes

$$\left(\frac{\partial u}{\partial t}\right)_{z=0} = -gZ. \qquad (4.17)$$

As it stands this condition contains the unknown free surface height Z. However, Z can be eliminated by using the kinematic condition that the normal components of the velocity of the surface itself, and of the fluid there, are equal. Since the disturbances are weak, it is sufficient to take the normal direction in the z-direction as in the undisturbed flow. Thus, we obtain

$$\frac{\partial Z}{\partial t} = \frac{\partial u}{\partial z}, \qquad (4.18)$$

to be applied on $z = 0$.

Both conditions (4.17) and (4.18) apply for all t and elimination of Z between them yields the boundary condition for u at the free surface,

$$\frac{\partial^2 u}{\partial t^2} + g \frac{\partial u}{\partial z} = 0, \quad z = 0, \quad \text{all } t.$$

For an application of this boundary condition see Exercise 4.3.

In the calculation of steady temperature distributions, Newton's Law of Cooling (Section 3.3) provides a general relation between the temperature and its normal derivative at each point on the boundary. As before, as special limiting cases, the temperature may be prescribed or have zero normal derivative.

Lastly, the same condition which gave rise to (4.14) on the velocity potential also requires the stream function U to be constant on solid boundaries at rest since such a boundary is a streamline.

4.4 Properties of Laplace's Equation

As the simplest representative of the class of elliptic partial differential equations, Laplace's equation has solutions which have fundamentally different properties from those of the equations treated in Chapters 2 and 3. However, in appearance it resembles the wave equation (2.1), and it is tempting to apply to (4.1) the type of analysis used in discussing (2.1) in Section 2.3. It should be understood that the fact that the ratio of the second derivatives has opposite signs in the two cases is crucial and the interpretation of (4.1) as a wave equation with imaginary 'wave speeds', i and $-i$, $(i = \sqrt{(-1)})$, in the (x, y) plane is without physical meaning. Nonetheless, proceeding tentatively, the analysis that led to the definition of the characteristic coordinates, $\xi = x - ct, \eta = x + ct$, in

Section 2.3 can be formally repeated here, where we write instead, $z = x + iy, \bar{z} = x - iy$. The rules for transforming derivatives to the new variables are

$$\frac{\partial}{\partial x} = \frac{\partial}{\partial z} + \frac{\partial}{\partial \bar{z}}, \quad \frac{\partial}{\partial y} = i\left(\frac{\partial}{\partial z} - \frac{\partial}{\partial \bar{z}}\right),$$

and conversely,

$$\frac{\partial}{\partial z} = \frac{1}{2}\left(\frac{\partial}{\partial x} - i\frac{\partial}{\partial y}\right), \quad \frac{\partial}{\partial \bar{z}} = \frac{1}{2}\left(\frac{\partial}{\partial x} + i\frac{\partial}{\partial y}\right).$$

Hence, the Laplacian operator becomes

$$\frac{\partial^2}{\partial x^2} + \frac{\partial^2}{\partial y^2} = \left(\frac{\partial}{\partial x} - i\frac{\partial}{\partial y}\right)\left(\frac{\partial}{\partial x} + i\frac{\partial}{\partial y}\right) = 4\frac{\partial^2}{\partial \bar{z}\,\partial z}. \tag{4.19}$$

We note that the characteristics z = const., $\bar{z}$ = const., are complex and represent no real lines in the (x, y) plane. This is a property of elliptic equations in general and can be used as the definition. Thus we expect that the role of characteristics in obtaining solutions, so useful in the case of hyperbolic equations, is not likely to be so fruitful in the elliptic case.

The simple form of the Laplacian (4.19), when expressed in the coordinates z and $\bar{z}$, suggests that the theory of functions of a complex variable plays an important part in the study of the two-dimensional Laplace's equation. This is, in fact, the case and whilst it is not the intention in this book to go deeply into methods of solution based on complex variable theory, it is worthwhile pursuing the matter a little further.

First we recall the result that a necessary and sufficient condition for the complex-valued function $\phi(x, y) + i\psi(x, y)$ to be an analytic function of $z = x + iy$ in a domain D (that is, it is differentiable in a neighbourhood of every point in D) is that ϕ and ψ are single-valued and continuous functions of x and y, together with their first partial derivatives, in D and that the Cauchy–Riemann equations,

$$\frac{\partial \phi}{\partial x} = \frac{\partial \psi}{\partial y}, \quad \frac{\partial \phi}{\partial y} = -\frac{\partial \psi}{\partial x}, \tag{4.20}$$

hold there.

Now, on writing $w = \phi + i\psi$, we may express these conditions in a concise way by transforming to the variable $\bar{z}$. Thus, we may write

$$0 = \frac{\partial \phi}{\partial x} - \frac{\partial \psi}{\partial y} + i\left(\frac{\partial \psi}{\partial x} + \frac{\partial \phi}{\partial y}\right) = \frac{\partial}{\partial x}(\phi + i\psi) + i\frac{\partial}{\partial y}(\phi + i\psi) = 2\frac{\partial w}{\partial \bar{z}}. \tag{4.21}$$

Equation (4.21) states that when any two real functions of x and y, say $\phi(x, y)$ and $\psi(x, y)$ (satisfying the continuity and differentiability conditions mentioned above), are combined together in the form $\phi + i\psi$ to produce a complex function w of z and $\bar{z}$, in general, then w is analytic only when it is independent of $\bar{z}$.

To see how this result may be applied to Laplace's equation we first deduce from (4.19) that a complex function w is a solution if $\partial w / \partial z$ is an analytic function of z. It follows then from a theorem in complex variable theory that w itself is analytic when calculated by integration along a path in any simply-connected domain in which $\partial w / \partial z$ is analytic. Thus the requirement that a complex function satisfies Laplace's equation is that it is analytic, and since the equation is linear the real and imaginary parts of such a function are harmonic. We stress that the foregoing process is a formal one with no rigorous justification for the introduction of the differential operators $\partial / \partial z$ and $\partial / \partial \bar{z}$, and was suggested by the corresponding process for the wave equation. However, we can easily verify directly that if ϕ and ψ satisfy (4.20) then they are harmonic.

This tie-up between the theory of harmonic functions and that of analytic functions of a complex variable can be used to derive important properties satisfied by the former. As an example, a well-known theorem in complex variable theory, the maximum modulus theorem, states that the modulus of any function that is analytic in a bounded domain and continuous on the boundary assumes its maximum value somewhere on the boundary of the domain and nowhere at an interior point, unless the function is identically a constant. The corresponding result holds for minimum values of the modulus if the additional condition, that the function nowhere vanishes in the domain or on its boundary, is satisfied. Now the function e^w is analytic wherever w is and its modulus is e^ϕ, which is an increasing function of ϕ and never vanishes. Hence if $\phi(x, y)$ is harmonic within some bounded domain and continuous on the boundary it assumes its maximum and minimum values on the boundary unless ϕ is a constant.

This approach establishes the result for Laplace's equation in two dimensions; it is also true when the number of independent variables is increased and, indeed, for much more general types of elliptic partial differential equations. As a physical illustration of the maximum and minimum principles for harmonic functions we note that in steady conditions it is impossible for maximum and minimum temperatures to occur in the interior of a medium unless the temperature is uniform throughout. Any such hot or cold 'spot' in the interior would create a flow of heat by the Second Law of Thermodynamics and the assumption of time-independence would be violated.

We now leave the complex variable approach to equation (4.1) and, as with the wave and diffusion equations, consider the associated difference equation. In this case, the latter is,

$$\Delta_{xx}u + \Delta_{yy}u = 0, \tag{4.22}$$

where Δ_{xx} is defined by equation (2.29) and Δ_{yy} by (2.30) on substituting y for t and for simplicity choosing $k = h$. From (4.22) we can obtain an expression for $u(x, y)$ in the form,

$$u(x, y) = \tfrac{1}{4}\{u(x + h, y) + u(x - h, y) + u(x, y + h) + u(x, y - h)\}. \tag{4.23}$$

Thus, as illustrated in Fig. 4.2, the value of u at any interior point of the region (one whose neighbours all lie in the region) is the average value of u at the four neighbouring mesh points. This is clearly connected with the property of the differential equation just discussed, that solutions possess no maximum (and minimum) in the interior because an average value at a point implies a greater (and a smaller) value at a neighbouring point unless u has the same value at all mesh points.

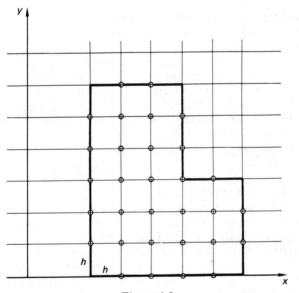

Figure 4.2

For the boundary value problem, with u given at every point of the boundary (a point at least one of whose neighbours lies outside the region and one within it), corresponding to the Dirichlet problem for the differential equation defined in Section 1.4, it is clear that a single equation can be written down for each internal point. If there are n such points then the n values of u, say u_n, at these points satisfy n simultaneous algebraic equations and these equations are linear. They therefore have a unique solution for u_n if the coefficient determinant is non-zero. That this is so may be readily seen for the special case where the boundary values are all zero. For then, if the determinant were to vanish, the equations for u_n would be homogeneous and have a solution other than the trivial solution $u_n = 0$, all n. Without loss of generality suppose such a solution has a positive maximum at some point x_i, y_i, where $u = u_i = M$, say. Then, in view of equation (4.23), at the four neighbouring mesh points the values of u must be M since none can be greater than M. The argument can then be repeated to show that at all the mesh points the values of u are M and this contradicts the assumption of zero boundary values. Finally, since the coefficient determinant is independent of boundary values it follows that a unique solution for the u_n exists whatever values are taken for u on the boundary. Note that all the boundary must be included; otherwise the system of n equations would contain more than n unknowns and the solution would not be unique.

4.5 Solutions of Laplace's Equation

In this section we apply the method of separation of variables to obtain solutions of Laplace's equation satisfying boundary conditions of the general types discussed in Section 4.3. When the geometry of the region in which we wish to obtain solutions is rectangular, as in example 4.5.1 below, the method follows closely that for the wave and diffusion equations. But it is often desirable to discuss more general shapes of boundaries and an illustration of this is provided by example 4.5.2 which deals with a circular boundary. It is then more convenient to work in plane polar coordinates and the Cartesian form of Laplace's equation (4.1) must first be transformed into these coordinates before the method can be applied.

4.5.1 Steady State Heat Flow in a Slab

Here a slab is taken to mean a long solid cylinder of rectangular cross-section with generators parallel to the z-direction. The cross-section in the (x, y) plane is drawn in Fig. 4.3 and we now consider the problem of calculating the steady state temperature distribution in the rectangle $0 \leqslant x \leqslant a, 0 \leqslant y \leqslant b$, when different types of boundary conditions are applied to the faces $x = 0, x = a, y = 0, y = b$. The problem treated in Section 3.5.1 is fundamentally different from the present one since we are there concerned with the unsteady one-dimensional flow of heat.

Example (i). The first specific problem we look at is that in which the faces $x = 0, x = a$ and $y = 0$ are all held at constant temperature which, without loss of generality, we can take to be zero, and a prescribed temperature distribution is maintained on the face $y = b$. Thus we consider a Dirichlet problem for the rectangle with

$$\theta = 0 \text{ on } x = 0 \quad \text{and} \quad x = a \quad \text{for } 0 \leqslant y \leqslant b, \qquad (4.24)$$

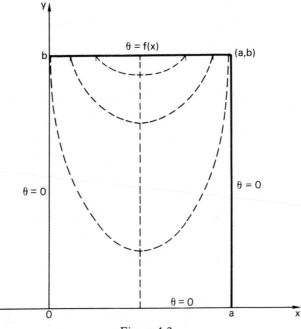

Figure 4.3

$\theta = 0$ on $y = 0$, $\theta = f(x)$ on $y = b$,   for $0 \leqslant x \leqslant a$,

$$\frac{\partial^2 \theta}{\partial x^2} + \frac{\partial^2 \theta}{\partial y^2} = 0 \text{ in } 0 \leqslant x \leqslant a, \quad 0 \leqslant y \leqslant b. \tag{4.25}$$

As usual we look for a solution of the form,

$$\theta(x, y) = X(x)Y(y). \tag{4.26}$$

Equation (4.25) is satisfied by (4.26) if

$$\frac{1}{X}\frac{d^2 X}{dx^2} = -\frac{1}{Y}\frac{d^2 Y}{dy^2},$$

and hence each side of the equation must be separately constant. Consideration of the boundary conditions on $x = 0$ and $x = a$, for reasons already discussed, quickly leads to the result that the constant must be negative, say $-m^2$. Then,

$$\frac{d^2 X}{dx^2} + m^2 X = 0, \quad \frac{d^2 Y}{dy^2} - m^2 Y = 0. \tag{4.27a, b}$$

The general solution of equation (4.27a) has already been written down (see equation (2.47)) and in exactly the same way as in the case of a vibrating string fixed at $x = 0$ and $x = l$ (Section 2.7) we find that $A = 0$ and $m = n\pi/a$ in equation (2.47) for all integral n. Hence the function,

$$X = B_n \sin\left(\frac{n\pi x}{a}\right), \tag{4.28}$$

for arbitrary B_n is the general solution of (4.27a) satisfying the boundary conditions (4.24).

The general solution of equation (4.27b) for Y, however, is different from any we have obtained previously. It is a linear equation with constant coefficients and a standard method, for example by D-operators, produces the simple result that

$$Y = C' e^{my} + D' e^{-my},$$

where C' and D' are arbitrary constants. It is often useful to write this result in the alternative form,

$$Y = \left(\frac{C' - D'}{2}\right)(e^{my} - e^{-my}) + \left(\frac{C' + D'}{2}\right)(e^{my} + e^{-my})$$

$$= C \sinh my + D \cosh my,$$

where $C = C' - D'$ and $D = C' + D'$ are further arbitrary constants. The full solution for $\theta(x, y)$ can then be represented by the infinite series,

$$\theta(x, y) = \sum_{n=1}^{\infty}\left[C_n \sinh\left(\frac{n\pi y}{a}\right) + D_n \cosh\left(\frac{n\pi y}{a}\right)\right]\sin\left(\frac{n\pi x}{a}\right), \tag{4.29}$$

on using (4.26), superposing the solutions for different values of n and absorbing the B_n in (4.28) into the arbitrary constants C_n and D_n. To choose C_n and D_n the boundary conditions on $y = 0$ and $y = b$ must be applied. Since $\theta = 0$ on $y = 0$,

$$\sum_{n=1}^{\infty} D_n \sin\left(\frac{n\pi x}{a}\right) = 0, \quad 0 \leqslant x \leqslant a,$$

and hence $D_n = 0$ for all n. The boundary condition on $y = b$ then requires that

$$f(x) = \sum_{n=1}^{\infty} C_n \sinh\left(\frac{n\pi b}{a}\right)\sin\left(\frac{n\pi x}{a}\right), \quad 0 \leqslant x \leqslant a. \tag{4.30}$$

The problem is thus reduced to the determination of the Fourier sine series for $f(x)$ and it is now clear how the solution depends on the particular form of the prescribed temperature maintained at the face $y = b$ of the slab. If a simple temperature distribution $f(x) = K \sin(\pi x/a)$, with K constant, were maintained there, for example by electrical means, the infinite series reduces to its first term only and $C_1 \sinh(\pi b/a) = K$, $C_n = 0$ ($n \neq 1$). The solution for any point of the slab is then

$$\theta(x, y) = K \sinh\left(\frac{\pi y}{a}\right) \sin\left(\frac{\pi x}{a}\right) \Big/ \sinh\left(\frac{\pi b}{a}\right). \qquad (4.31)$$

More generally the constants C_n are given, by (1.27), as

$$C_n = \frac{2}{a \sinh(n\pi b/a)} \int_0^a f(x) \sin\left(\frac{n\pi x}{a}\right) dx,$$

and the resulting solution is then

$$\theta(x, y) = \sum_{n=1}^{\infty} C_n \sinh\left(\frac{n\pi y}{a}\right) \sin\left(\frac{n\pi x}{a}\right). \qquad (4.32)$$

For example, if the surface $y = b$ were maintained at a uniform temperature so that $f(x) = $ constant, say θ_0, the coefficients C_n are

$$C_n = \frac{2\theta_0}{a \sinh(n\pi b/a)} \int_0^a \sin\left(\frac{n\pi x}{a}\right) dx = \frac{2\theta_0[1 - (-1)^n]}{n\pi \sinh(n\pi b/a)},$$

and hence, the temperature at any point in the slab is given by

$$\theta(x, y) = \frac{2\theta_0}{\pi} \sum_{n=1}^{\infty} \frac{[1 - (-1)^n]}{n \sinh(n\pi b/a)} \sinh\left(\frac{n\pi y}{a}\right) \sin\left(\frac{n\pi x}{a}\right). \qquad (4.33)$$

On the other hand if the temperature distribution on $y = b$ were symmetric and parabolic we could write $f(x) = 4\theta_0 x(a - x)/a^2$, where θ_0 is the maximum temperature. The constants C_n are then

$$C_n = \frac{8\theta_0}{a^3 \sinh(n\pi b/a)} \int_0^a x(a - x) \sin\left(\frac{n\pi x}{a}\right) dx$$

$$= \frac{8\theta_0}{a^2 n\pi \sinh(n\pi b/a)} \int_0^a (a - 2x) \cos\left(\frac{n\pi x}{a}\right) dx$$

$$= \frac{16\theta_0}{a(n\pi)^2 \sinh(n\pi b/a)} \int_0^a \sin\left(\frac{n\pi x}{a}\right) dx$$

$$= \frac{16\theta_0[1 - (-1)^n]}{(n\pi)^3 \sinh(n\pi b/a)},$$

on integrating twice by parts.

Hence

$$\theta(x, y) = \frac{16\theta_0}{\pi^3} \sum_{n=1}^{\infty} \frac{[1 - (-1)^n] \sinh(n\pi y/a) \sin(n\pi x/a)}{n^3 \sinh(n\pi b/a)}. \quad (4.34)$$

We see that the coefficients in this series are $1/n^2$ times a constant multiple of the corresponding coefficients in the first case, (4.33), and the series therefore converges much more rapidly. This is connected with the fact that the sum of the Fourier series outside the domain of definition of $f(x)$, $0 < x < a$, being the odd, periodic extension of $f(x)$, is continuous for the parabolic temperature distribution but not for the uniform one.

Incidentally, the first terms in the series (4.33) and (4.34) differ only by about 3%. This is a consequence of the fact that the applied sinusoidal and parabolic distributions, having the same maximum θ_0, are in reasonable agreement all along the face $y = b$.

From these solutions we can deduce some results of physical interest on the behaviour of the temperature. To simplify the discussion we confine it to the case of the sinusoidal temperature distribution on $y = b$ for which the solution is (4.31). The surfaces of constant temperature, or isothermals, in the slab are given by $\theta(x, y) = $ constant and if we specify the constant by means of the value x_1, at which they intersect the plane $y = b$, the equation for the family of isothermals is

$$\sinh\left(\frac{\pi y}{a}\right) \sin\left(\frac{\pi x}{a}\right) = \sinh\left(\frac{\pi b}{a}\right) \sin\left(\frac{\pi x_1}{a}\right), \quad 0 \leqslant x_1 \leqslant a, \quad (4.35)$$

for which the domains of x and y are $0 \leqslant x \leqslant a$, $0 \leqslant y \leqslant b$. These surfaces are symmetric about the plane $x = a/2$ as we should expect from the physics of the problem since the imposed boundary conditions are symmetric with respect to this plane. Sketches of the isothermals are given in Fig. 4.3.

A quantity of physical importance is the rate at which heat flows across the boundary at $y = b$ in order to keep the temperature there at its prescribed value, since this rate is proportional to the power requirement of the external generating mechanism. This rate of heat flow per unit area on $y = b$ is proportional to

$$-\left(\frac{\partial\theta}{\partial y}\right)_{y=b} = -\frac{K\pi}{a} \coth\left(\frac{\pi b}{a}\right) \sin\left(\frac{\pi x}{a}\right), \quad 0 \leqslant x \leqslant a.$$

We see that this is negative implying that the flow of heat is in the $-y$-direction, that is into the slab, and it is this heat which must be supplied by external means if the steady temperature state in the slab is to be maintained. In a similar way we can calculate the flow of heat per unit area across the other faces of the slab; apart from a multiplicative constant these are

$$-\left(\frac{\partial\theta}{\partial y}\right)_{y=0} = -\frac{K\pi \sin\left(\dfrac{\pi x}{a}\right)}{a \sinh\left(\dfrac{\pi b}{a}\right)},$$

$$-\left(\frac{\partial\theta}{\partial x}\right)_{x=0} = +\left(\frac{\partial\theta}{\partial x}\right)_{x=a} = -\frac{K\pi \sinh\left(\dfrac{\pi y}{a}\right)}{a \sinh\left(\dfrac{\pi b}{a}\right)}.$$

From these expressions it is readily confirmed that heat flows out of the slab over the faces $y = 0, x = 0$ and $x = a$.

Formula (3.2) gives the rate of heat flow per unit area across a plane. By considering planes parallel to the (y, z) and (x, z) planes in turn we can introduce the idea of a heat flow vector, say $\mathbf{J}(x, y)$, which represents the heat flow at a point in magnitude and direction in terms of its components in the $\mathbf{i}$ and $\mathbf{j}$ directions. Thus

$$\mathbf{J} = -k\left(\frac{\partial\theta}{\partial x}\mathbf{i} + \frac{\partial\theta}{\partial y}\mathbf{j}\right) = -k \operatorname{grad} \theta,$$

in two dimensions, on using (3.2). A family of surfaces whose slope at any point is the direction of $\mathbf{J}$ can thus be drawn and is defined by

$$\frac{dy}{dx} = \frac{\partial\theta/\partial y}{\partial\theta/\partial x} = \coth\left(\frac{\pi y}{a}\right)\tan\left(\frac{\pi x}{a}\right), \tag{4.36}$$

for the special case we are here considering. Since, from (4.31), the slope of the isothermal through (x, y) is

$$\frac{dy}{dx} = -\tanh\left(\frac{\pi y}{a}\right)\cot\left(\frac{\pi x}{a}\right), \tag{4.37}$$

the product of the two values of dy/dx in (4.36) and (4.37) is -1 and the two families of surfaces are orthogonal. This result is generally true (Exercise 4.14).

Finally, in connection with the particular solution (4.31), we note that the maximum temperature it represents is K, which is not attained in the interior of the slab, and therefore the maximum principle is verified. Similarly, there is no point of zero temperature within the slab.

This heat flow problem in the slab is mathematically identical with the problem of finding the electrostatic potential in a vacuum bounded by a cylinder of rectangular cross-section with thin metallic walls, three of which, the planes $x = 0, x = a$ and $y = 0$, are earthed and separated by insulating strips from the fourth, $y = b$, which is maintained at a potential

$f(x)$. The solution for the potential corresponds to that for the temperature, (4.32), and the surfaces in the same direction as the electric intensity, or flux surfaces, correspond to those in the direction of the heat flow vector defined above. This interpretation of the solution of Laplace's equation has application in electronics for calculating the potential inside a vacuum tube.

Example (ii).   As a second heat flow problem in the slab we suppose that the surface, $y = 0$, is insulated instead of being maintained at the fixed temperature $\theta = 0$. Thus the boundary conditions are now (4.24) and

$$\frac{\partial \theta}{\partial y} = 0 \text{ on } y = 0, \quad \theta = f(x) \text{ on } y = b, \quad \text{for } 0 \leqslant x \leqslant a. \qquad (4.38)$$

There is no difference from the analysis in Example (i) as far as the general solution (4.29). However, on applying the boundary conditions (4.38) we see that now $C_n = 0$ for all n and the constants D_n are simply related to the coefficients of the Fourier sine series for $f(x)$. They satisfy

$$f(x) = \sum_{n=1}^{\infty} D_n \cosh\left(\frac{n\pi b}{a}\right) \sin\left(\frac{n\pi x}{a}\right),$$

and on comparing this series with (4.30) we deduce that for a given temperature distribution $f(x)$ on $y = b$ the constants required in the solution for the two problems are related by $D_n = C_n \tanh(n\pi b/a)$. We can therefore write down the temperature distribution within the slab for the three particular choices made for $f(x)$ in Example (i).

First, for the simple temperature distribution $f(x) = K \sin(\pi x/a)$, we obtain

$$\theta(x, y) = K \cosh(\pi y/a) \sin(\pi x/a)/\cosh(\pi b/a).$$

Thus, there is here a sinusoidal distribution of temperature, $K \operatorname{sech}(\pi b/a)$ $\sin(\pi x/a)$ on the face $y = 0$. It is easy to write down an equation, analogous to (4.35), for the family of isothermals. These curves have the same general shape as those drawn in Fig. 4.3 but the isothermal passing through a given point (x, y) has slope of greater magnitude than in the previous case. This may be understood when we recognize that heat, which flows at right angles to the isothermals, can escape only through the faces $x = 0$ and $x = a$ and not through the insulated face $y = 0$. In fact, although there is a smaller heat inflow per unit time across $y = b$ in this case there is a greater outflow per unit time across $x = 0$ and $x = a$. These results are left for the reader to verify for himself (also see Exercise 4.9).

For the case where $f(x) = \theta_0$, constant, the solution is, from (4.33),

$$\theta(x, y) = \frac{2\theta_0}{\pi} \sum_{n=1}^{\infty} \frac{[1 - (-1)^n] \cosh(n\pi y/a) \sin(n\pi x/a)}{n \cosh(n\pi b/a)},$$

and when $f(x) = 4\theta_0 x(a - x)/a^2$, from (4.34),

$$\theta(x, y) = \frac{16\theta_0}{\pi^3} \sum_{n=1}^{\infty} \frac{[1 - (-1)^n] \cosh(n\pi y/a) \sin(n\pi x/a)}{n^3 \cosh(n\pi b/a)}.$$

Again, we note that these series and their counterparts in Example (i), (4.33) and (4.34), contain only sines of odd multiples of $\pi x/a$. Thus the temperature in the slab is symmetric about the plane $x = a/2$ and this reflects the corresponding symmetry of the applied boundary conditions in each case.

Before leaving the heat flow problem in the slab it is instructive to consider other boundary conditions that might be applied. The more general Newtonian cooling condition in the simplest situation where the medium in $y < 0$ is at zero temperature becomes $\partial\theta/\partial y = -h\theta$ on $y = 0$, from (3.8), and then C_n and D_n in (4.29) are proportional for all n. On the other hand, if the face $y = 0$ has a prescribed, maintained, non-zero temperature distribution, $\theta = g(x)$ say, Fourier sine series are needed for both $f(x)$ and $g(x)$ to determine C_n and D_n.

By relaxing the zero temperature conditions on the side walls $x = 0$ and $x = a$ a further range of problems is obtained. The simplest of these is the heat-insulated case with $\partial\theta/\partial x = 0$ on $x = 0$ and $x = a$, when Fourier cosine series (corresponding to $m = n\pi/a$, n an integer) rather than sine series are appropriate, whilst, if one wall is at zero temperature and the other insulated, Fourier series corresponding to $m = (2n + 1)\pi/2$ are required analogous to Example (ii) of Section 2.7. In the more general cases involving the application of Newton's Law of Cooling on $x = 0$ or $x = a$, m is found to satisfy a transcendental equation of the type mentioned at the end of Section 3.5.1.

4.5.2 Problems in Plane Polar Coordinates

When the boundary in the (x, y) plane of the region in which solutions of Laplace's equation are sought is circular it is more convenient to use plane polar coordinates (r, ϕ). The relations between the two systems of coordinates are

$$x = r \cos \phi, \quad y = r \sin \phi,$$

or, in inverse form,

$$r = \sqrt{(x^2 + y^2)}, \quad \phi = \tan^{-1}(y/x), \quad -\pi < \phi \leqslant \pi.$$

The transformation of the differential operators is carried out by means of the Chain Rule (cf. Section 2.3) and hence

$$\frac{\partial}{\partial x} = \cos \phi \frac{\partial}{\partial r} - \frac{\sin \phi}{r} \frac{\partial}{\partial \phi}, \quad \frac{\partial}{\partial y} = \sin \phi \frac{\partial}{\partial r} + \frac{\cos \phi}{r} \frac{\partial}{\partial \phi}.$$

On applying this transformation again we obtain Laplace's equation in plane polar coordinates,

$$\frac{\partial^2 u}{\partial x^2} + \frac{\partial^2 u}{\partial y^2} = \frac{\partial^2 u}{\partial r^2} + \frac{1}{r} \frac{\partial u}{\partial r} + \frac{1}{r^2} \frac{\partial^2 u}{\partial \phi^2} = 0. \tag{4.39}$$

We now suppose that the boundary of the region within which we are to solve (4.39) is $r = a$ and that conditions governing the existence of a solution independent of the third space coordinate z are satisfied (see, for example, Section 4.2.1). The boundary conditions must also be expressed in terms of r and ϕ. Thus, for the case of a specified value of u on $r = a$ (Dirichlet problem) it takes the form $u(a, \phi) = f(\phi), -\pi < \phi \leqslant \pi$, ($f$ given), whereas if the normal derivative is given all round the boundary (Neumann problem) we have $\partial u/\partial r = g(\phi), r = a, -\pi < \phi \leqslant \pi$, for some specified g. Clearly both these conditions may apply simultaneously over smaller, mutually exclusive ranges of ϕ, or, again, other boundary conditions such as the type arising from Newton's Law of Cooling may be appropriate. Even then problems may be subdivided into further classes; interior problems where the region of interest is $r < a$ and exterior problems where this region is $r > a$. We shall discuss examples taken from each of these classes. Although the treatment is similar there are some points of difference which are of interest.

(i) Interior Dirichlet Problem for a Circle.   In this case we must satisfy (4.39) in $r < a$ subject to $u(a, \phi) = f(\phi), -\pi < \phi \leqslant \pi$, where f is a known function. Physically we can think of a temperature distribution prescribed and maintained on the surface of a long, cylindrical rod or electrostatic potential given on a long, conducting, cylindrical shell; the solution of (4.39) in $r < a$ then represents the steady temperature distribution in the rod or potential in the vacuum bounded by the shell, respectively.

We apply the method of separation of variables to (4.39) so we assume a solution of the form,

$$u(r, \phi) = R(r)\Phi(\phi).$$

On differentiating and substituting into (4.39) we obtain an equation in R and Φ which can be arranged as

$$\frac{1}{R}\left(r^2 \frac{d^2 R}{dr^2} + r \frac{dR}{dr}\right) = -\frac{1}{\Phi} \frac{d^2 \Phi}{d\phi^2}.$$

The reason for writing the equation in this way is so that the standard argument in the method can be applied. Thus, the left-hand side is independent of ϕ and the right-hand side independent of r and, since the equation holds for all r and ϕ in $0 \leqslant r < a$, $-\pi < \phi \leqslant \pi$, each side must be equal to the same constant, say m^2. The equations for R and Φ therefore become

$$r^2 \frac{d^2R}{dr^2} + r \frac{dR}{dr} - m^2 R = 0, \quad \frac{d^2\Phi}{d\phi^2} + m^2\Phi = 0. \tag{4.40a, b}$$

We have not hitherto met an equation of the type (4.40a) but it has the property of being homogeneous in r; changing r to λr (λ constant) has no effect on the equation. Since differentiation with respect to r reduces a power of r by one and the powers of r occurring in (4.40a) multiply derivatives of the same order, a simple power of r is suggested as a solution. Hence, we substitute $R = r^N$ into (4.40a) and deduce that the two possible values of N are $\pm m$ with $m > 0$. The solution is therefore

$$R = Ar^m + Br^{-m}, \quad m > 0, \tag{4.41}$$

where A and B are arbitrary constants.

When the separation constant, m, equals zero, (4.40a) can be written in the form $d(r\,dR/dr)/dr = 0$, which has the solution,

$$R = A + B \log r. \tag{4.42}$$

From the physical problem, the temperature or potential must be finite along the axis of the cylinder, $r = 0$, and hence $B = 0$ in the above expressions.

The solution of (4.40b) is

$$\Phi = C \cos m\phi + D \sin m\phi, \tag{4.43}$$

but a further physical requirement is that the solution should be a single-valued function of position within the cylinder. This implies that Φ is periodic in ϕ with period 2π. Thus, $\Phi(\phi + 2\pi) = \Phi(\phi)$, and it follows that m is an integer n which is positive or zero. On combining the resulting solution with (4.41) and applying the Principle of Superposition, we obtain the more general solution,

$$u(r, \phi) = \sum_{n=0}^{\infty} r^n (C_n \cos n\phi + D_n \sin n\phi), \quad r \leqslant a. \tag{4.44}$$

To determine the coefficients C_n and D_n we apply the boundary condition that $u = f(\phi)$ on $r = a$. Therefore,

$$f(\phi) = \sum_{n=0}^{\infty} a^n (C_n \cos n\phi + D_n \sin n\phi), \quad -\pi < \phi \leqslant \pi,$$

and, hence, C_n and D_n are the coefficients of the Fourier series for $f(\phi)$ and from (1.23) are given by

$$
\left.
\begin{array}{l}
C_0 = \dfrac{1}{2\pi} \displaystyle\int_{-\pi}^{\pi} f(\phi)\, d\phi, \\[3mm]
a^n C_n = \dfrac{1}{\pi} \displaystyle\int_{-\pi}^{\pi} f(\phi)\cos n\phi\, d\phi,\quad a^n D_n = \dfrac{1}{\pi} \displaystyle\int_{-\pi}^{\pi} f(\phi)\sin n\phi\, d\phi, \\[3mm]
n = 1, 2, 3, \ldots..
\end{array}
\right\} \quad (4.45)
$$

With C_n and D_n given by these expressions, the solution is now known as a series, (4.44). It can in fact be expressed as an integral (see Exercise 4.12) known as Poisson's formula for a circle.

Notice from (4.44) that the value of u at the centre of the circle is $u(0, \phi) = C_0$, which is given by (4.45). This provides a general result for harmonic functions. Since the position of the circle and its radius may be chosen arbitrarily, a function $u(x, y)$ at a point P is equal to its average value taken over any circular boundary with centre P within which u is harmonic. In particular, for example, the temperature along the axis of the cylinder is equal to the average temperature over the circumference.

(ii) Exterior Problem for a Circle.   A simple convenient example of the exterior problem for a circle occurs when a solid, circular cylinder is placed in an infinite volume of incompressible fluid which, far from the cylinder, is moving with uniform velocity normal to its axis. Although, in practice, effects of viscosity, leading to the formation of a substantial wake behind the cylinder, are significant in such a flow we assume the fluid to be inviscid and the flow to be irrotational. The solution is still useful, as mentioned below. In Section 4.2.2 we saw that the velocity potential u satisfies Laplace's equation, in this case for the region $r > a$, and from (4.14) the normal component of the velocity on the boundary must vanish;

$$
\nabla u \cdot \mathbf{n} = \frac{\partial u}{\partial n} = \frac{\partial u}{\partial r} = 0 \quad \text{on } r = a. \tag{4.46}
$$

In addition we assume that the disturbance created by the cylinder in the fluid dies away far from the cylinder itself. Thus, as $r \to \infty$, the potential approaches the value for a uniform velocity, say parallel to the x-direction (Fig. 4.4). Then $\partial u/\partial x = U$, say, $\partial u/\partial y = 0$, for large r and therefore $u \to Ux$ if we ignore any constant in u. The condition to be placed on u is therefore

$$
u \to Ur \cos \phi \quad \text{as } r \to \infty. \tag{4.47}
$$

The solution by separation of variables proceeds in the same way as for the interior problem. The periodicity condition again implies that Φ is given by (4.43) with m a non-negative integer. By using expressions (4.41) and (4.42) for R a series solution is obtained by superposition as

$$u = A + B \log r + \sum_{n=1}^{\infty} \{(C_n' \cos n\phi + D_n' \sin n\phi)r^n +$$

$$+ (C_n'' \cos n\phi + D_n'' \sin n\phi)r^{-n}\} \tag{4.48}$$

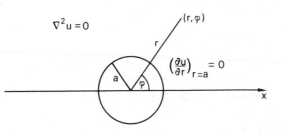

Figure 4.4

where we have introduced further arbitrary constants C_n', D_n', C_n'' and D_n''. Application of the boundary condition at infinity, (4.47), implies that the only positive power of r that can be present in (4.48) is the first and the constant and logarithmic terms must also be omitted. Hence, $C_1' = U$, $A = B = 0$, $C_n' = 0 \ (n \neq 1)$, $D_n' = 0 \ (n \geq 1)$. The resulting series for $\partial u/\partial r$ is

$$\frac{\partial u}{\partial r} = U \cos \phi - \sum_{n=1}^{\infty} nr^{-n-1}(C_n'' \cos n\phi + D_n'' \sin n\phi),$$

and the boundary condition (4.46) on the cylinder's surface determines the constants C_n'' and D_n'' as $C_1'' = Ua^2$, $C_n'' = 0 \ (n \neq 1)$, $D_n'' = 0 \ (n \geq 1)$. Thus, in this case, the infinite series becomes a finite one of two terms,

$$u = U\left(r + \frac{a^2}{r}\right)\cos \phi, \quad r \geq a, \quad -\pi < \phi \leq \pi. \tag{4.49}$$

This result is of limited applicability to the actual flow past a circular cylinder because of the effects of the viscosity of the fluid. Its importance lies in the fact that flows past streamlined shapes, such as those suitable for aeroplane wing cross-sections and for which adverse effects of the fluid's viscosity are greatly reduced, can be found from (4.49) by the application of conformal transformations in complex variable theory to map the shape onto the circle. We shall not pursue this aspect here but merely mention that in the applications to aerofoil design the more general solution of the exterior problem for the circle, which allows an arbitrary circulation of velocity round the circle (as well as the uniform flow past it), is often needed. The necessary modification to the foregoing problem is the inclusion of an extra condition,

$$\oint_C \nabla u . ds = \Gamma,$$

where Γ is the given circulation and we may take C, a closed curve encircling the cylinder, to be a concentric circle to it. Thus, for points on C,

$$\Gamma = u(r, \phi + 2\pi) - u(r, \phi), \tag{4.50}$$

and the potential is no longer periodic although, of course, the velocity must be. The latter requirement means that m must again be an integer but to incorporate the condition (4.50) we must include the full solution of (4.40b) when $m = 0$, namely,

$$\Phi = C_0 + D_0\phi.$$

Then, the general solution is effectively (4.49) plus an additional term, $(A + B \log r)\phi$. The condition at $r = \infty$ requires that $B = 0$ and the constant A is determined from the condition (4.50).

EXERCISES

1.   Verify that the vector field $-Km\mathbf{r}/r^3$ is both irrotational and solenoidal.

2.   The theory of inviscid, irrotational, subsonic flow past a thin body, under suitable conditions, is governed by the equation $\partial^2\phi/\partial y^2 + \beta^2\partial^2\phi/\partial x^2 = 0$ for the velocity potential ϕ of the disturbance, where the constant β satisfies $1 \geqslant \beta > 0$ and $\beta = 1$ for incompressible flow (cf. Exercise 2.4 for the supersonic case). Show that the transformation $\phi(x, y) = k_1\Phi(X, Y), x = X, y = k_2 Y$, for particular constant values of k_1 and k_2, reduces this equation to Laplace's equation for $\Phi(X, Y)$. The transformation thus relates a given compressible flow, ϕ, to an incompressible one, Φ. By using $dy/dx \propto \partial\phi/\partial y$ as the boundary condition at the surface choose k_1 such that ϕ and Φ represent flows past the same body. Show that the effect of compressibility is to increase pressure differences, and hence lift, by a factor β^{-1}.

3.   Water, of undisturbed depth h, is contained in a tank of rectangular cross-section and which is long in the x-direction. Show that a solution of Laplace's equation representing the velocity potential for a wave

travelling in the x-direction is $u = (A\,e^{mz} + B\,e^{-mz})\cos[m(x - ct)]$, where m and c are wave number and wave speed and A and B are constants. By applying the boundary conditions at the bottom and the free surface find the relation between A and B and show that

$$c^2 = \frac{g}{2\pi}\,\lambda\,\tanh\left(\frac{2\pi h}{\lambda}\right),$$

where λ is the wavelength.

4. Show that the fluid speed cannot have a maximum in the interior of an incompressible fluid. (Hint. For an arbitrary internal point P take the x-direction in the direction of the fluid velocity at P.)

5. Solve the difference equation $(\Delta_{xx} + \Delta_{yy})u = 0$ for u at the points (h, h), $(2h, h)$ and $(3h, h)$ if the boundary values are $u(x, 0) = T$, $u(x, 2h) = 2T$ and $u(0, h) = u(4h, h) = 0$. Verify that at these points $0 < u < 2T$.

6. The simple equation (1.20) is the one-dimensional form of Laplace's equation. If θ denotes the temperature in a material of thermal conductivity k, find the heat loss per unit area across faces at $x = 0$ and $x = l$ which are kept at temperatures of θ_d and θ_1 respectively. Deduce the effect of increasing (i) $\theta_1 - \theta_0$, (ii) k, (iii) l.

7. Apply the result obtained in Exercise 5 to estimate the heat loss through the walls, windows, roof and floor of a house of cubic shape with side of 10 m in conditions in which the inside is kept at $20°C$. above the outside, if the values of k per unit thickness per unit area are 1.5, 4.5, 2.0 and 1.5 Watts per $°C$. per m^2, respectively, for the walls, windows, roof and floor. Deduce the benefit of roof insulation and double-glazing if the heat loss through roof and windows is thereby reduced by a factor two. (Assume the windows take up 40% of the vertical area.)

8. As in Section 4.5.1 find the steady state heat flow in a slab whose faces $x = 0$, $x = a$ and $y = 0$ are insulated, and $y = b$ is kept at a temperature (i) $\theta_0\cos(\pi x/a)$, (ii) $\theta_0\sin(\pi x/a)$.

9. Using the solution (4.31) verify that the total heat inflow into the slab vanishes, in keeping with the existence of a steady state.

10. Find the electrostatic potential u in a vacuum bounded by thin metallic walls, insulated from each other, in the shape of a long rectangle such that the boundary conditions may be taken as $u(0, y) = u(a, y) = 0$, $u(x, y) \to 0$ as $y \to \infty$ and $u(x, 0) = x$, $0 \leqslant x \leqslant a/2$, $u(x, 0) = a - x$, $a/2 \leqslant x \leqslant a$.

11.   Classify the types of series required for the solution of Laplace's equation in the rectangle $(0, 0)$, $(a, 0)$, (a, b), $(0, b)$, when boundary conditions of the Dirichlet, Neumann or general type (3.8) are imposed on the various sides.

12.   (i) Substitute the expressions (4.45) for C_n and D_n into the series (4.44) and derive Poisson's formula for the solution of Laplace's equation inside the circle $r = a$,

$$u(r, \phi) = \frac{1}{2\pi} \int_{-\pi}^{\pi} f(\psi) \left\{ \frac{a^2 - r^2}{a^2 - 2ar \cos(\phi - \psi) + r^2} \right\} d\psi, \qquad (4.51)$$

where ψ is a variable of integration.

(ii)   Following closely the analysis in Section 4.5.2(a), derive the Poisson formula, analogous to (4.51), for the exterior Dirichlet problem to the circle $r = a$ if the solution u satisfies $u = f(\phi)$ on $r = a$ and, in addition, $u \to 0$ as $r \to \infty$. Give a physical interpretation of this problem.

13.   Find the steady state temperature distribution in a long annular tube whose inner surface of radius a is maintained at a temperature of $\theta_0 \cos^2 \phi$ and outer surface of radius b is insulated.

14.   Prove in general the result verified for the special case discussed in Section 4.5.1, that the isothermals are orthogonal to the family of surfaces in the direction of $\mathbf{J}$.

15.   Derive the inviscid, irrotational flow past a circular cylinder, discussed in Section 4.5.2(b), by the alternative formulation using the stream function defined in Section 4.2.3.

16.   The potential distribution on the surface of a long circular cylinder of radius a is $u = V_0\phi^2$, $-\pi < \phi \leqslant \pi$. Calculate the potential at any point of the vacuum region outside $r = a$, assuming that it remains bounded as r tends to infinity.

Further Reading

The subject of partial differential equations is vast and at the present time occupies the attention of large numbers of mathematicians. They are concerned not only with the solution of equations under specified boundary conditions, by many different methods, but also with the more theoretical questions of existence, uniqueness and properties of solutions. As a first introduction to the subject, throughout this book we have touched on just a few of these topics in their most elementary form. It would be out of place here to attempt to give a comprehensive bibliography, even if that were possible, but for those students who are interested in pursuing the subject in greater depth or breadth a list of

selected reading is appended. In the first group the books include
material at roughly the same level, although possibly with a different
emphasis from that adopted here; in the second the books are generally
much more advanced and, in some cases, have become standard works.

Bland, D. R., *Solutions of Laplace's Equation*, 1961 (Routledge and
  Kegan Paul).
Bland, D. R., *Vibrating Strings*, 1960 (Routledge and Kegan Paul).
Churchill, R. V., *Fourier Series and Boundary Value Problems*, 1963
  (McGraw-Hill).
Coulson, C. A., *Waves*, 1952 (Oliver and Boyd).
Moon, P. and Spencer, D. E., *Partial Differential Equations*, 1969
  (Heath and Company).
Sneddon, I. N., *Elements of Partial Differential Equations*, 1957
  (McGraw-Hill).
Stephenson, G., *An Introduction to Partial Differential Equations for
  Science Students,* 1970 (Longman).

Bateman, H., *Partial Differential Equations of Mathematical Physics*,
  1964 (Cambridge University Press).
Carslaw, H. S. and Jaeger, J. C., *Conduction of Heat in Solids*, 1959
  (Oxford University Press).
Courant, R. and Hilbert, D., *Methods of Mathematical Physics II: Partial
  Differential Equations*, 1962 (Interscience).
Fox, L., *Numerical Solution of Ordinary and Partial Differential Equations*,
  1962 (Pergamon Press).
Garabedian, P. R., *Partial Differential Equations*, 1964 (John Wiley and
  Sons).
Jeffreys, H. and Jeffreys, B. S., *Methods of Mathematical Physics*, 1966
  (Cambridge University Press).
Kellogg, O. D., *Foundations of Potential Theory*, 1953 (Dover
  Publications).
Morse, P. M. and Feshbach, H., *Methods of Theoretical Physics*, 1953
  (McGraw-Hill).

Answers

CHAPTER 1

4. $ab > 0$, hyperbolic for $|x| < |hy|\sqrt{(ab)}$, parabolic for $|x| = |hy|\sqrt{(ab)}$, elliptic for $|x| > |hy|\sqrt{(ab)}$. $ab \leqslant 0$, hyperbolic for x or $y \neq 0$.

11. (a) $\dfrac{1}{4} + \displaystyle\sum_{n=1}^{\infty} \left\{ \dfrac{[1-(-1)^n]}{(n\pi)^2} \cos n\pi x + \dfrac{1}{n\pi} \sin n\pi x \right\}$,

　(b) $\displaystyle\sum_{n=1}^{\infty} \dfrac{2}{n\pi} \sin n\pi x$

　(c) $\dfrac{1}{2} + \displaystyle\sum_{n=1}^{\infty} \dfrac{2[1-(-1)^n]}{(n\pi)^2} \cos n\pi x$,

　$\dfrac{1}{2}, 0, 1.$

12. $\dfrac{4}{\pi}\left\{ \dfrac{1}{2} - \displaystyle\sum_{r=1}^{\infty} \dfrac{1}{(4r^2 - 1)} \cos\left(\dfrac{2r\pi x}{1} \right) \right\}, -\sin\left(\dfrac{\pi x}{1} \right).$

CHAPTER 2

1. (i) $u = 0$, (ii) $\dfrac{\partial u}{\partial x} = 0$, (iii) $M\dfrac{\partial^2 u}{\partial t^2} = -E\dfrac{\partial u}{\partial x}$, all evaluated at the end.

4. For $y > 0$, $\phi = -\zeta_1(x - \beta y)/\beta$; for $y < 0$, $\phi = \zeta_2(x + \beta y)/\beta$.

5. $t \leqslant b/c$: $x < -b - ct$, $u = 0$; $-b - ct < x < -b + ct$,
   $u = a[1 - (x + ct)^2/b^2]/2$; $-b + ct < x < b - ct$,
   $u = a[2 - (x + ct)^2/b^2 - (x - ct)^2/b^2]/2$; $b - ct < x < b + ct$,
   $u = a[1 - (x - ct)^2/b^2]/2$; $x > b + ct$, $u = 0$.
   $t \geqslant b/c$: $x < -b - ct$, $u = 0$; $-b - ct < x < b - ct$,
   $u = a[1 - (x + ct)^2/b^2]/2$; $b - ct < x < -b + ct$, $u = 0$;
   $-b + ct < x < b + ct$, $u = a[1 - (x - ct)^2/b^2]/2$; $x > b + ct$, $u = 0$.

6. For same five regions as in Exercise 5.
   $t \leqslant b/c$: $0, (x + ct + b)/2c, Ut, (b - x + ct)/2c, 0$.
   $t \geqslant b/c$: $0, (x + ct + b)/2c, Ub/c, (b - x + ct)/2c, 0$.

7. $P \sin [(x + ct)/b]$.

9. $Z = 0.05 \sin [4\pi(x - t)]$, 0.5 s, 2 s.$^{-1}$, 2 m.$^{-1}$, -0.05 m., 0 ms.$^{-1}$

10. $|A_r| = A_i\pi mnc/T_0 D$, $|A_t| = A_i/D$, $D = \sqrt{(1 + \pi^2 m^2 n^2 c^2/T_0^2)}$.
    $\mathrm{Arg}\, A_r = -\pi/2 - \epsilon$, $\mathrm{Arg}\, A_t = -\epsilon$, $\epsilon = \tan^{-1}(\pi mnc/T_0)$, $0 \leqslant \epsilon < \pi/2$.

12. $(S_2 - S_1)^2/(S_2 + S_1)^2, 4S_1S_2/(S_2 + S_1)^2.$

13. $|A_r| = A_ikc/4\pi nT_0 D, |A_t| = A_i/D, D = \sqrt{(1 + k^2c^2/16\pi^2n^2T_0^2)}.$
    $\text{Arg } A_r = \pi/2 + \epsilon, \text{Arg } A_t = \epsilon, \epsilon = \tan^{-1}(kc/4\pi nT_0), 0 \leqslant \epsilon < \pi/2.$

16. $a \cos\left(\dfrac{\pi ct}{l}\right) \sin\left(\dfrac{\pi x}{l}\right) + \dfrac{b}{2\pi} \sin\left(\dfrac{2\pi ct}{l}\right) \sin\left(\dfrac{2\pi x}{l}\right).$

17. $\dfrac{4Ul}{\pi^2 c} \displaystyle\sum_{r=0}^{\infty} \dfrac{(-1)^r}{(2r+1)^2} \sin\left[\dfrac{(2r+1)\pi}{8}\right] \sin\left[\dfrac{(2r+1)\pi x}{l}\right] \sin\left[\dfrac{(2r+1)\pi ct}{l}\right].$

18. (i) $\bar{p} = \dfrac{4K}{\pi} \displaystyle\sum_{r=0}^{\infty} \dfrac{1}{(2r+1)} \sin\left[\dfrac{(2r+1)\pi x}{l}\right] \cos\left[\dfrac{(2r+1)\pi ct}{l}\right]$, (ii) $v = \dfrac{V\bar{p}}{K}$,
    $\bar{p}$ in part (i).

19. $\dfrac{8\alpha}{\pi^2} \displaystyle\sum_{n=0}^{\infty} \dfrac{(-1)^n}{(2n+1)^2} \sin\left[\dfrac{(2n+1)\pi x}{2l}\right] \cos\left[\dfrac{(2n+1)\pi ct}{2l}\right].$

21. (i) 9, (ii) $17 \pm 12\sqrt{2}.$

22. $\rho_0 c^2 a^2 \pi^2 k^2 \sin^2 2\pi kct, T_0 a^2 \pi^2 k^2 \cos^2 2\pi kct.$

23. $\rho_0 U^2 l/8.$

24. $\rho_0 V^2 l/2, 8/\pi^2; K\alpha^2/2l, 8/\pi^2.$

25. $\rho ga^2 l/8 + \dfrac{\rho ga^2 l}{\pi^2} \displaystyle\sum_{r=0}^{\infty} \dfrac{1}{(2r+1)^2} \cos^2\left[\dfrac{(2r+1)\pi ct}{l}\right];$
    $\dfrac{\rho g^2 ha^2 l}{\pi^2 c^2} \displaystyle\sum_{r=0}^{\infty} \dfrac{1}{(2r+1)^2} \sin^2\left[\dfrac{(2r+1)\pi ct}{l}\right].$

CHAPTER 3

5. $\displaystyle\sum_{n=0}^{\infty} \dfrac{16l^2}{[(2n+1)\pi]^3} [(2n+1)\pi(-1)^n - 2] \sin\left[\dfrac{(2n+1)\pi x}{2l}\right]$
   $\exp\left[-\alpha^2(2n+1)^2\pi^2 t/4l^2\right].$

6. $\theta_1 + \dfrac{2}{\pi}(\theta_0 - 2\theta_1) \displaystyle\sum_{r=0}^{\infty} \dfrac{1}{(2r+1)} \sin\left[\dfrac{(2r+1)\pi x}{l}\right]$
   $\exp\left[-\alpha^2(2r+1)^2\pi^2 t/l^2\right] - \dfrac{4\theta_0}{\pi} \displaystyle\sum_{n=1}^{\infty} \dfrac{r}{(4r^2-1)} \sin\left(\dfrac{2r\pi x}{l}\right)$
   $\exp(-\alpha^2 4r^2\pi^2 t/l^2).$

7. $\theta_1 + \dfrac{2}{\pi} \displaystyle\sum_{n=0}^{\infty} (-1)^n \left\{\dfrac{\theta_0 - 2\theta_1}{2n+1} + \dfrac{(2n+1)\theta_0}{(2n-1)(2n+3)}\right\} \cos\left[\dfrac{(2n+1)\pi x}{2l}\right]$
   $\exp\left[-\alpha^2(2n+1)\pi^2 t/4l^2\right].$

8. $\dfrac{V}{Rl} + \dfrac{2V}{Rl} \displaystyle\sum_{n=1}^{\infty} \cos\left(\dfrac{n\pi x}{l}\right) \exp(-n^2\pi^2\alpha^2 t/l^2).$

9. $\dfrac{l^2}{6} - \displaystyle\sum_{r=1}^{\infty} \dfrac{l^2}{\pi^2 r^2} \cos\left(\dfrac{2r\pi x}{l}\right) \exp\left(-4\alpha^2 r^2 \pi^2 t/l^2\right).$

10. $-V_0 \exp\left(-x^2/4\alpha^2 t\right)/\alpha R \sqrt{(\pi t)}.$

11. $U\{1 - \text{erf}[x/2\sqrt{(\nu t)}]\}.$

12. Frictional force per unit area is $-\rho U \sqrt{\left(\dfrac{n\nu}{2}\right)}$ [sinh $2\Omega l$(cos nt

   $-$ sin nt) + sin $2\Omega l$(cos nt + sin nt)] /(cosh $2\Omega l$ $-$ cos $2\Omega l$), $\Omega = \sqrt{(n/2\nu)}.$

13. $\rho U \sqrt{(n\nu/2)}$ [sin Ωl(cos nt + sin nt) $-$ sinh Ωl(cos nt $-$ sin nt)] /
   (cosh Ωl + cos Ωl).

14. $\alpha\sqrt{(2/n)}, \theta_1 \exp\left[-\sqrt{(n/2\alpha^2)}x\right].$

15. $\theta_0 + 2\theta_1 \left[e^{-\Omega_1 l} \cos(\Omega_1 l + nt) + e^{\Omega_1 l} \cos(\Omega_1 l - nt)\right]/2(\cosh 2\Omega_1 l$
   $+ \cos 2\Omega_1 l), \Omega = \sqrt{(n/2\alpha^2)}.$

CHAPTER 4

6. $-k(\theta_1 - \theta_0)/l.$

7. 32% reduction.

8. (i) $\theta_0 \cosh(\pi y/a) \cos(\pi x/a)/\cosh(\pi b/a)$, (ii) $(2\theta_0/\pi)\{1 - \displaystyle\sum_{r=1}^{\infty}$
   $[2 \cosh(2\pi r y/a) \cos(2\pi r x/a)/(4r^2 - 1) \cosh(2\pi r b/a)]\}.$

10. $\displaystyle\sum_{r=0}^{\infty} \dfrac{(-1)^r 4a}{[\pi(2r+1)]^2} \exp\left[-(2r+1)\pi y/a\right] \sin\left[(2r+1)\pi x/a\right].$

13. $\dfrac{\theta_0}{2}\left[1 + \dfrac{a^2(r^4 + b^4)}{r^2(a^4 + b^4)} \cos 2\phi\right].$

16. $\dfrac{V_0 \pi^2}{3} + \displaystyle\sum_{n=1}^{\infty} \dfrac{4V_0(-1)^n}{n^2} \cos n\phi.$

Index

117